ECOCRITICISM AND THE IDEA OF CULTURE

for Esther and Harold

Ecocriticism and the Idea of Culture

Biology and the Bildungsroman

HELENA FEDER

East Carolina University, USA

LONDON AND NEW YORK

First published 2014 by Ashgate Publishing

Published 2016 by Routledge
2 Park Square, Milton Park, Abingdon, Oxfordshire OX14 4RN
711 Third Avenue, New York, NY 10017, USA

First issued in paperback 2016

Routledge is an imprint of the Taylor & Francis Group, an informa business

British Library Cataloguing in Publication Data
A catalogue record for this book is available from the British Library

The Library of Congress has cataloged the printed edition as follows:
Feder, Helena.
  Ecocriticism and the idea of culture: biology and the bildungsroman / by Helena Feder.
    pages cm
  Includes index.
  ISBN 978-1-4094-0157-5 (hardcover: alk. paper)

  1. Bildungsromans—History and criticism. 2. Ecology in literature. 3. Nature in literature
I. Title.
  PN3448.B54F43 2013
  809.3'9353—dc23

                                                            2013026671

ISBN 13: 978-1-138-24985-1 (pbk)
ISBN 13: 978-1-4094-0157-5 (hbk)

Contents

Chapter 1
Introduction:
Biology and the Idea of Culture

> There is so much resistance to the idea of animal culture that one cannot escape
> the impression that it is an idea whose time has come.
> —Frans de Waal, *The Ape and the Sushi Master*

The idea of nature has long been the subject of ecocritical analysis. Ecological thinkers have amply demonstrated the dangers of a notion of nature that excludes culture and its role in ecological crisis; it positions human beings as outside ecological conditions and superior to the other inhabitants of the world. However, the idea of culture defined by this binary, the exclusive realm of human enterprise, has not been adequately considered.

When ecocritical work has discussed culture *as such* in the last decade and a half, it has often been in the process of contesting a view of nature as a cultural construction. As ecocritics have pointed out, though this constructionist view of nature seems to "undo" the binary of nature and culture, it often merely replaces one side of the equation with the other. Taken to its extreme, this paradigm denies the cogent reality of materiality, of an agential world apart from human culture. Yet, as postmodern, poststructuralist, Marxist, and other theorists have pointed out, isn't everything "always already" mediated by culture? In this tired debate, ecocritics, busy refuting an erasure of nature, and other theorists, busy asserting the primacy of culture, both end up affirming the essentialist idea of culture at the core of this binary and the humanities. The persistence of this formulation of culture is *the* most pressing philosophical problem for ecocriticism and green studies, and critical and cultural theory generally.

In 1980, Lewis Thomas expressed frustration with cultural criticism's fascination with physics, especially quantum mechanics. "I wish the humanists," he wrote, "would leave physics alone for a while and begin paying more attention to biology" (70). The need for a more biologically, ecologically informed critique is, if anything, now more urgent.[1] By turning to biology, cultural biology, and related branches of the life sciences, we find the broader and more nuanced notion of culture necessary for a materialist ecocritical practice. While our experience of the

---

[1]   Glen Love's *Practical Ecocriticism* (2003) and numerous articles and conference panels since its publication call for greater scientific awareness in ecocritical practice. Echoing Thomas, Love remarks, "If some humanists have been attracted to some of the most difficult and obscure physics, they have for the most part ignored the life sciences, especially evolutionary biology and ecology" (49).

world is culturally mediated and constructed, culture is itself a product of nature, and human culture is only one of many types of culture in the material world.

In light of this expansive notion of culture, discussed in this chapter, this book considers one of the most enduring of modern Western cultural forms, the Bildungsroman, not only as the novel of individual development but also as humanism's origin story of culture. This "ecocultural materialist" approach to various examples of genre, including François-Marie Arouet de Voltaire's *Candide,* Mary Shelley's *Frankenstein*, Virginia Woolf's *Orlando*, and Jamaica Kincaid's *A Small Place*, reveals the foundational opposition of "nature" and "culture" as a tension that sometimes manifests itself as anxiety, sometimes as marked fluidity, sometimes as inversion. In these radical examples of the genre— or examples read radically—this tension suggests, however latent or denied, humanism's knowledge of nonhuman agency and, sometimes, subjectivity. If critique is to intervene meaningfully in our historical crises, it must move beyond solipsistic (that is, solely anthropological) notions of society and culture. The purpose of attempting a broader scope for Marxist cultural analysis and a dialectical methodology for ecocriticism is not to engage debates about the nature of ideology or anthropocentrism, but to suggest a necessarily more diverse, complex field for *materialist* critiques that already tend to analyze systematic rationalism (industries, institutions, discourses, etc.) in terms of the domination of human and nonhuman nature.

Recognizing the existence of other animal cultures—and, in so doing, rejecting various ideologies of nature, particularly that of human supremacy—challenges structures of power that oppress both human and nonhuman animals. My project here is twofold: to consider the Bildungsroman in terms of humanism's claim about our radical uniqueness, to see how examples of the genre reveal the cracks at the core of this claim, and to work toward an ecocultural materialism. It is at once an experiment in "immanent critique," an examination of the form and content of ideology in the service of emancipatory knowledge,[2] and a participant in the recent

---

[2]   As Terry Eagleton explains in *Ideology*, "it is perfectly possible, as with the Marxist concept of an 'immanent' critique, to launch a radical critique of culture from somewhere inside it, not least from those internal fissures or fault-lines which betray its underlying contradictions" (4). In "Cultural Criticism and Society," Theodor Adorno acknowledges that while dialectical critique is always in a sense both transcendent and immanent, immanent criticism is the more inherently dialectical of the two modes of analysis. "The choice of a standpoint outside the sway of existing society is as fictitious as only the construction of abstract utopias can be. Hence, the transcendent criticism of culture, much like bourgeois cultural criticism, sees itself obliged to fall back upon the idea of 'naturalness,' which itself forms a central element of bourgeois ideology. The transcendent attack on culture regularly speaks the language of false escape, that of the 'nature boy.' ... Against this struggles the immanent procedure as the more essentially dialectical ... The traditional transcendent critique of ideology is obsolete" (*Prisms* 31–3).

materialist turn in theory.[3] A more materialist, more "worldly" multiculturalism might intervene in forms of oppression that have long functioned by excluding some—human and nonhuman—from the realm of culture. This question of culture is, of course, not only a disciplinary but a political one. And, in the end, the very idea of politics—politics itself—is what is at stake.

Disciplinary/Politics

In May of 2010, the United Nations International Year of Biodiversity, geneticist Craig Venter and his research team created what he calls "the world's first synthetic life form"—a bacterium described as "a defining moment in biology." Venter claims this single-celled organism with its made-from-scratch genome "heralds the dawn of a new era in which new life is made to benefit humanity, starting with bacteria that churn out biofuels, soak up carbon dioxide from the atmosphere and even manufacture vaccines" (Sample, "Craig Venter").[4] This new lifeform, invention and intervention, is a source of tremendous interest and anxiety—not unlike Alan Weisman's *The World Without Us*, "a penetrating, page-turning tour of a post-human Earth," twenty-six weeks on *The New York Times* bestseller list, *Time Magazine*'s number one nonfiction book of 2007, and inspiration for the 2010 television series (and IPhone App), *Life After People*.[5] This novel creature's

---

[3] For example, in 2010 Stacy Alaimo describes this turn with respect to feminist theory: "What has been most notably excluded by the 'primacy of cultural' and the turn toward the linguistic and the discursive is the 'stuff' of matter. Theorists within the overlapping fields of feminist theory, environmental theory, and science studies, however, have put forth innovative understandings of the material world. Some feminist theorists, such as Moira Gatens, Claire Colebrook, and Elizabeth Bray, have embraced the work of Spinoza and Deleuze as counter traditions to the linguistic turn. Others have reread theorists at the heart of poststructuralism—for example, Jacques Derrida (Vicki Kirby and Elizabeth Wilson), Michel Foucault (Ladelle McWhorter and Karen Barad), and Judith Butler (Karen Barad). Together, these theorists, along with others, constitute the materialist turn in feminist theory" (*Bodily Natures* 6).

[4] "To mark the genome as synthetic, they spliced in fresh strands of DNA, each a biological 'watermark' that would do nothing in the final organism except carry coded messages, including a line from James Joyce: 'To live, to err, to fall, to triumph, to recreate life out of life'" (Sample, "Synthentic").

[5] See http://www.worldwithoutus.com/about_book.html. *Life After People* is a series on the History Channel: "In every episode, viewers will witness the epic destruction of iconic structures and buildings, from the Sears Tower, Astrodome, and Chrysler Building to the Sistine Chapel ... With humans gone, animals will inherit the places where we once lived. Elephants that escape from the LA Zoo will thrive in a region once dominated by their ancestors, the wooly mammoth. Alligators will move into sub-tropical cities like Houston—feeding off household pets. Tens of thousands of hogs, domesticated for food, will flourish. In a world without people, new stories of predators, survival and evolution will emerge. Humans won't be around forever, and now we can see in detail, for the very first time, the world that will be left behind in *Life After People: The Series*." See http://www.history.com/shows/life-after-people/articles/about-life-after-people.

place in material reality and, conversely, the resonance of a world after humanity in the human imagination would seem to confirm what some theorists have argued for decades, that at some point we or the world became "posthuman." Increasingly, when we encounter the subject of biology in the media or in scholarship in the humanities, it is in the context of posthumanism.

All of which begs the question: is it we *or* the world, or we *as* the world? We imagine life after people not only because human sovereignty over the rest of life on earth continues to intensify exponentially, but also because our cultures (perhaps particularly, but not only, in the West) tell us we *are* the world, the pinnacle or "brain" of nature, nature's self-reflexive agency or *natura naturans*[6] (even as we render it less and less inhabitable for ourselves and many other creatures). If evolutionary thought and the ecological sciences have taught nothing else, surely it has taught us that humans are not *the* world. And yet we are the world too—our bodies are themselves ecosystems, our atoms the very fibers of it. In *Bodily Natures*, Stacy Alaimo argues that we are not so much corporeal as "trans-corporeal": "the human is always inter-meshed with the more-than-human world" (2).[7] But while we and the world interpenetrate, we do not equate. As D.H. Lawrence wrote of Whitman's pantheism, "Aristotle did not live for nothing. All Walt is Pan, but all Pan is not Walt" (*Phoenix* 24). While we are still here—the world is not yet without us even if it is too much with us—the apocalyptic *post* of posthumanism warns that the Age of "Man" may soon give way to an age without human beings (at least, as we have known them[8]) and a great many others.

While the term posthumanism may reinforce ideas of human supremacy, promoting life beyond biology (after all, we are the creators of wholly new life

---

6   For two assertions of *natura naturans*, both of which posit problematic readings of *The Winter's Tale*, see Frederick W. Turner's "Cultivating the American Garden" in *The Ecocriticism Reader* and Terry Eagleton's *The Idea of Culture*. Far better, though equally problematic, Eagleton argues: "We resemble nature in that we, like it, are to be cuffed into shape, but we differ from it in that we can do this to ourselves, thus introducing into the world a degree of self-reflexivity to which the rest of nature cannot aspire. ... Cultivation, however, may not only be something we do to ourselves. It may also be something done to us, not least by the political state. For the state to flourish, it must inculcate in its citizens the proper sorts of spiritual disposition; and it is this which the idea of culture or *Bildung* signifies in a venerable tradition from Schiller to Mathew Arnold" (6).

7   Alaimo's project here overlaps with but is very different from mine. She is one of the very few (if not the only) scholar working in the field at the moment to state specifically that nonhuman animals have culture: "Rather than arguing, however, that humans are natural creatures, that nonhuman animals are cultural creatures, and that the nature/culture divide is not sustainable (all of which I believe), I will locate my inquiry within the many interfaces between human bodies and the larger environment" (4).

8   As Rob Nixon recently wrote, we are faced with "amorphous calamities," not only in the form of the Anthropocene (and its "Great Acceleration" of CO_2 emissions since 1950) but also in the age of the new "man"—in which "high speed planetary modification has been accompanied (at least for those increasing billions who have access to the Internet) by rapid modifications to the human cortex"; these increasing connections, and "the paradoxical disconnects that can accompany it," seem to redefine time itself (12).

forms "made to benefit humanity") and sometimes, more specifically, human life beyond the current bounds of humanity (cue the robotic and virtualized selves of the imagination), it also signals a renewed interest in the biological world, human animality, and our kinship with other creatures (as we see in the field of animal studies).[9] Posthumanism may challenge the primacy of humanity or it may champion a humanist teleology, a race for infinite technological power over material life; it may function as a landscape of virtuality or a deeper recognition of the connections between material agencies, a reimagining of what Darwin described in *Origin of the Species* as "a web of complex relations." In short, posthumanism may mean many things, some of which are mutually-exclusive: a revaluing of human animality or the desire to transcend animality; a radical, ecological sensibility; or a teleological essentialism.

While humanism is certainly far more complex than any caricature of the Enlightenment[10] (as I will argue in the next chapter), there is no mistaking its essentialist legacy. In *What is Posthumanism?* Cary Wolfe also differentiates between the two poles of the term, between transhumanist (teleological, transcendent) and critical (materialist) posthumanisms, but argues that even critical posthumanism must move beyond "a thematic of the decentering of the human" to challenge the form of thought itself if it is to be truly "posthuman." In the field of animal studies, the radical impact of posthumanism ("what makes it not just another flavor of 'fill in the blank' studies") "is that it fundamentally unsettles and reconfigures the question of the knowing subject and the disciplinary paradigms and procedures that take for granted its form and reproduce it." Wolfe argues that the posthuman challenge of this field is lost when "the animal" becomes simply another "object" of study (xxix).

Like animal studies, ecocriticism is in the process of contesting paradigms and considering conditions of knowledge as well as the purposes of such knowledge. We too must focus on our philosophical, disciplinary challenge to the anthropocentric orthodoxies of the humanities. Ecocriticism's radical challenge lies not only in recognizing other forms of subjectivity and the ecological interconnectedness of biologically diverse subjects, but in recognizing that the relations between them are *political*—they are life and death relations. We are one animal among many in this shared world, living in interwoven interspecies communities, a series of polises themselves comprised of differing societies. This is not to say that this political work must take the form of human political relations, or that the ethical

---

9   In *How We Became Posthuman*, N. Katherine Hayles asserts the duality of what is called posthumanism: the rejection or erasure of the body or materiality for a fantasy of disembodiment *and* the realization of that fantasy's root in the familiar subject of liberal humanism, with its disavowal of embodiment and embeddedness in pursuit of individuality and freedom. This realization makes possible the second posthumanism, the critique that reveals that the human of humanism, the free-floating Cartesian mind, or the atomized subject of "free" political-economy, is itself a fantasy. This posthumanism suggests that, far from finding ourselves on the far side of an historic rupture, we may have always been posthuman, even as it offers new modes of subjectivity (Hayles 2).

10   A point Neil Badmington makes in "Theorizing Posthumanism."

consideration of other animals[11] depends on how "intelligent" or *like us* we think they are, but that we must begin to take seriously the implications of our real similarities with and differences from other creatures. As Terry Eagleton famously argued, "Political argument is not an alternative to moral preoccupations: it is those preoccupations taken seriously in their full implications" (*Literary Theory* 208).

The discussion of politics is, of course, always itself political. And as Jacques Rancière suggests, what is at stake is the definition of politics itself:

> "Disagreement" and "dissensus" do not imply that politics is a struggle between camps; they imply that it is a struggle about what politics is, a struggle that is waged about such original issues as: "where are we?", "who are we?", "What makes us a we", "what do we see and what can we say about it that makes us a we, having a world in common?" Those paradoxical, unthinkable objects of thinking mark ... the places where the question 'How is this thinkable at all?' points to the question: "who is qualified for thinking at all?" (116)

We are all part of a common world, but one that is changing rapidly for the immediate benefit of some at the expense of a great many others. In this context, to ask who is qualified for politics, what counts as political, is to ask who *counts* full stop. For humanism (and, indeed, its uncritical *posts-*), the question of who counts is intimately bound up with the question of what counts as culture; to think politically, to think about politics, we must contest the humanist ideology of culture still at the core of the humanities and Western culture. To do this, we must look beyond laboratory cages, computers, and cyborgs. A more "worldly" critique requires a turn to the larger world.

Cultural Biology

> [I]f nature is dynamic and active, if it is not alien to culture but is the ground which makes the cultural logically and historically possible, then what would a new conception of culture, one which refuses to sever it from nature, look like?
>   —Elizabeth Grosz, *Time Travels*

While some scientists continue to disagree over the use of the term "culture,"[12] *Nature* and other prominent journals have published the findings of dozens of

---

[11]   As Derrida demonstrates in *The Animal That Therefore I Am,* there is no simple, wholly positive way to refer nonhuman animals; his partial solution to this problem is the term "animot." I will use several terms interchangeably—other animals, nonhuman animals, and animal-others—as these remind us of our animality and the reality of our political relations with the larger animal world—that is, their position as Other in our (and their) world.

[12]   There are skeptics, chief among them psychologist Bennett G. Galef, co-editor with Kevin N. Laland of *The Question of Animal Culture* (discussed in this chapter). In the humanities, neither animal studies nor ecocriticism seem to have processed the idea of nonhuman cultures, though an interest in biological research is rapidly growing. Love's *Practical Ecocriticism* stresses that "[b]iological evolution and cultural evolution

studies demonstrating that many species learn socially and pass on traditions or skills. For example, a comprehensive synthesis of several long-term studies of chimpanzees in Africa (151 years cumulatively) documents thirty-nine group-specific, learned behavioral patterns (including tool usage): "[T]he combined repertoire of these behavioral patterns in each chimpanzee community is itself highly distinctive, a phenomenon characteristic of human cultures but previously unrecognized in non-human species" (Whiten et al. 682). A particularly resonant example of learned tool use was reported in 2007 by researchers in Senegal, who recorded twenty-two examples of chimps creating spears to hunt smaller primates ("Chimpanzees 'hunt using spears'").[13] Primates, though, are not the only culture-makers in nature; evidence of animal cultures abounds—from Hal Whitehead's work on orcas and sperm whales to Kevin Laland's studies of birds and fish.[14] Writing on animal cultures, primatologist Frans de Waal exclaimed, "one cannot

---

are not independent but interrelated; hence such scientists' descriptions of the process as 'coevolutionary' or 'biocultural'" (19). Though Love does not discuss nonhuman cultures, he does note that the "traditional reluctance of many scientists and philosophers to attribute consciousness to animals must be questioned in the face of new evidence" (33). More recently, in "Eluding Capture: The Science, Culture, and Pleasure of 'Queer' Animals," Alaimo wrote, "Nonhuman animals are also cultural creatures, with their own sometimes complex systems of (often nonreproductive) sex. ... Rather than continuing to pose nature/culture dualisms that closet queer animals as well as animal cultures ... we can think of queer desire as part of an emergent universe of a multitude of naturecultures" (57–60).

[13] "Researchers documented 22 cases of chimps fashioning tools to jab at smaller primates sheltering in cavities of hollow branches or tree trunks. The report's authors, Jill Pruetz and Paco Bertolani, said the finding could have implications for human evolution. Chimps had not been previously observed hunting other animals with tools" (BBC).

[14] For example, see "Culture in Whales and Dolphins," by Luke Rendell and Hal Whitehead in *Behavioral and Brain Sciences*, and Whitehead's *Sperm Whales: Social Evolution in the Ocean*. On fish and birds, see Kevin N. Laland and William Hoppitt's "Do Animals Have Culture?" in *Evolutionary Anthropology*. While they take issue with the famous example of the potato-washing macaques, they do claim that some birds, whales, and fish have culture: "Cultures are those group-typical behavior patterns shared by members of a community that rely on socially learned and transmitted information. ... According to the preceding definition, which animals have culture? There are two kinds of answers to this question. The first kind is based exclusively on hard experimental evidence. That is, for which species do we have reliable scientific evidence of natural communities that share group-typical behavior patterns that are dependent on socially learned and transmitted information? The answer, which will surprise many, is humans plus a handful of species of birds, one or two whales, and two species of fish" (150–1). Also, see the recent issue *Culture Evolves* (edited by Andrew Whiten, Robert A. Hinde, Christopher B. Stringer and Kevin N. Laland), and John M. Marzluff and Tony Angell's *In the Company of Crows and Ravens*, which includes "a detailed look at the cultural life of crows, exploring their behavior and traditions and our influences on them." http://yalepress.yale.edu/yupbooks/excerpts/crows_and_ravens.asp.

escape the impression that it is an idea whose time has come" (13–14).[15] It is also an idea that has been kicking around, even if only to be dismissed, for quite some time.

In *Civilization and its Discontents*, Sigmund Freud considers, albeit briefly, the existence of nonhuman cultures:

> Why do our relatives, the animals, not exhibit any such cultural struggle? We do not know. Very probably some of them—the bees, the ants, the termites—strove for thousands of years before they arrived at the State institutions, the distribution of functions and the restrictions on the individual, for which we admire them today ... In the case of other animal species it may be that a temporary balance has been reached between the influences of their environment and the mutually contending instincts within them, and that thus a cessation of development has come about. (83)

Freud's question about animal culture was turned on its head (or, more accurately, stood on its feet) in 1953 when Kinji Imanishi, founder of Japanese primatology, applied ethnographic study to an animal society on the island of Koshima, creating animal cultural studies. In September of that year, Satsue Mito noticed Imo, an 18-month old macaque, carry a sweet potato to a freshwater stream and clean it before eating, minimizing wear on her teeth.[16]

> She playfully repeated this behavior on the first day. Later, she improved her technique by going deeper in the water, holding the potato in one hand and rubbing off the mud with the other, occasionally dipping it in the water ... Within three months, two of [Imo's] peers as well as her mother were showing the same behavior. From these potato pioneers the habit spread to other juveniles, their older siblings, and their mothers. Within five years, more than three quarters of the juveniles and young adults engaged in regular potato washing. (de Waal 200–201)

---

[15]   Even the Animal Planet network has a webpage on animal culture. Here is a sample from their five-page overview: "Primates are not the only animals in which scientists have discovered evidence of cultural transmission of behavior. Researchers believe the best nonprimate evidence for culture is found in songbirds, which include thrushes, jays, wrens, warblers, finches, and other common backyard birds. Many studies have indicated that songbirds learn their melodies from parents and neighbors of the same species. Songs within a particular species show regional variations similar to the regional dialects (variant forms of speech) common in human populations. ... [B]iologists think of the songs as culture because they represent behaviors that are transmitted through learning and imitation rather than being genetically determined." http://animals.howstuffworks.com/animal-facts/animal-culture-info.htm.

[16]   Imanishi concluded that the advantage to washing potatoes is the wear it saves on teeth. While Satsue Mito first observed and reported this behavior, Imanishi interpreted the behavior and his team conducted the formal research confirming social transmission.

This has become a rather famous example[17] of the "struggle" Freud did not see in the animal world: cultural change through socially learned problem solving.[18]

Building on the work of William McGrew's *Chimpanzee Material Culture* in 1992, primatologist Frans de Waal's *The Ape and the Sushi Master*, published in 2001, surveys and theorizes the methodological and conceptual issues of the growing field of animal cultural research, termed "cultural biology" (267).[19] He argues,

> The standard notion of humanity as the only form of life to have made the step from the natural to the cultural realm—as if one day we opened a door to a brand-new life—is in urgent need of correction ... The idea that we are the only species whose survival depends on culture is false, and the entire juxtaposing of nature and culture rests on a giant misunderstanding. (28)

De Waal goes on to state that even aesthetics may be found in nonhuman cultures: "Given that our aesthetic sense has been shaped by the environment in which we evolved, it is logical to expect preferences for shapes, contrasts, and colors to transcend species" (36).

While the question of aesthetics, and its associations with "high" culture, need not come into play here, it was just this view of culture that was used to deny its existence in various human groups. To my mind, incorporating the insights of cultural biology into analysis in the humanities is a clear continuation of the work of cultural materialism. As Tony Bennett reminds us in *New Keywords*, "By showing how the supposedly universal standards of perfection associated with the normative view of culture turned out, in practice, to have strong connections with the particular views of ruling groups and classes, [Raymond Williams] extended our sense of what might count as culture" (67). Let's return to Williams's foundational observation in *Keywords* that "culture" is "one of the two or three most complicated words in the English language ... [The Latin root] *Colere* has

---

[17] See Sara Shettleworth's *Cognition, Evolution, and Behavior* for a skeptical reading of this famous evidence. Also, see Galef's well-known 1990 article, "The Question of Animal Culture" in *Human Nature*. De Waal discusses Galef's argument in *The Ape and the Sushi Master*. "Galef questioned whether the spreading of potato washing had anything to do with imitation. The Canadian psychologist was right to take a close look at the evidence and to insist that scientists carefully weigh the options when they see a behavior spreading in a population. ... But given Galef's valid warning, it was all the more disturbing that he himself made so little effort to verify his own assumptions, for example, by actually visiting the island in person" (207).

[18] Potato washing, however, is not the only example of socially learned behavior on Koshima Island. "In 1956, she [Imo] introduced a solution to the problem that wheat thrown on to the beach mingles with sand. Imo learned to separate the two by carrying handfuls of the mixture to nearby water, and throwing it into it. Sand sinks faster than wheat, making for easy picking. This sluicing technique, too, was eventually adopted by most monkeys on the island" (de Waal 202).

[19] De Waal notes that this term was first proposed by Imanishi in 1950 (381).

a range of meanings: inhabit, cultivate, protect, honor with worship" (87). While all animal species inhabit, many live and learn socially, and some cultivate or transform food (such as leaf cutter ants). Even abstractions such as honor form a part of the lives of some animals. The elephant practice of ritual mourning is one such example.[20]

In 2005, Gay Bradshaw and her colleagues argued in *Nature* that human interference (poaching, "culling," and habitat loss) has led to a collapse of "elephant culture" (807). Wild elephants are demonstrating unprecedented aggression toward humans, and occasionally other animals, attacking villages and crops, killing hundreds of people each year. In an interview with Charles Siebert, Bradshaw describes this wide-scale phenomenon as psychological and cultural breakdown: "What we are seeing today is extraordinary. Where for centuries humans and elephants lived in relatively peaceful coexistence, there is now hostility and violence. Now, I use the term 'violence' because of the intentionality associated with it … " She asks, "How do we respond to the fact that we are causing other species like elephants to … breakdown? In a way, it's not so much a cognitive or imaginative leap anymore as it is a political one" ("An Elephant"). In *Elephants on the Edge*, Bradshaw contextualizes the implications of her research, interpreting elephant violence as another form of resistance to colonial oppression and global power:

> Much like other cultures that have refused to be absorbed by colonialism, elephants are struggling to survive as an intact society, to retain their elephant-ness, and to resist becoming what modern humanity has tried to make them—passive objects in zoos, circuses, and safari rides, romantic decorations dotting the landscape for eager eyes peering from Land Rovers, or data to tantalize our minds and stock in the bank of knowledge. Elephants are, as Archbishop Desmond Tutu wrote about black South Africans living under apartheid, simply asking to live in the land of their birth, where their dignity is acknowledged and respected. (71–2)

Bradshaw's work not only requires the recognition of our relations with elephants (and many other lifeforms) as political, it also suggests that the resistance to the idea of nonhuman animal cultures is not, or not only, intellectual but ideological. With many animals, including most mammals, and their habitats still treated as raw materials for production (much in the way other colonial subjects have been subject to horrific exploitation, physical and cultural genocide), the existence of other animal cultures, their numbers and scope, and the new political terrain they imply, present a profound challenge to power, including scientific humanism.

---

[20] For example, see work by Cynthia Moss, including "African Elephants Show High Levels of Interest in the Skulls and Ivory of Their Own Species" in *Biology Letters* and *Elephant Memories*. Also, see Marc Bekoff's *The Emotional Lives of Animals*. Finally, see Derrida's comment on this phenomenon in *The Animal That Therefore I Am*.

Carel van Schaik's *Among Orangutans: Red Apes and the Rise of Human Culture*, which documents twenty-four cultural variants among the orangutans he observed in Sumatra (including sophisticated tool-making and a variety of other socially learned behaviors), lays out the philosophical and scientific problem with traditional definitions of culture and the new biocentric corrective:

> The anthropological definitions emphasize the underlying beliefs and values of culture bearers ... The Japanese primatologist Kinji Imanishi was perhaps the first, in 1952, to point out that at its core, culture is socially transmitted innovation: culture is simply innovation followed by diffusion. This biological (as opposed to anthropological) definition leads to an operational emphasis on observable behaviors or artifacts, things we can actually see in animals, rather than beliefs or values, which we cannot. It also explains the key property of culture in humans: geographic variation. Useful or popular innovations spread until they hit some barrier, producing geographic differentiation. So, if we see geographic variation in behaviors that we know reflect innovation and are transmitted through some socially mediated learning process, then we have animal culture (and we can worry about how symbolic any of it is later on). (139)

However, in *Sense and Nonsense*, Kevin N. Laland and Gillian R. Brown assert that scientists are a long way from a consensus definition: "Most social scientists would agree on two points, that culture is composed of symbolically encoded acquired information and that it is socially transmitted within and between populations, largely free of biological constraints. Is that the way evolutionists regard culture? For the most part it would seem not" (310).[21] Put most simply, our notion of culture is culturally (and, more narrowly, disciplinarily) constructed; the emphasis on a narrow notion of symbol, along with symbolic learning and syntactic communication, is only one of the anthropological biases underlying some definitions of culture.

The definition of culture de Waal uses is as follows:

> Culture is a way of life shared by the members of one group but not necessarily with the member of other groups of the same species. It covers knowledge, habits, and skills, including underlying tendencies and preferences, derived from exposure to and learning from others. Whenever systematic variation in knowledge, habits, and skills between groups cannot be attributed to genetic or ecological factors, it is probably cultural. The way individuals learn from each other is secondary, but that they learn from each other is a requirement. (31)

Within the parameters of this definition, de Waal and other biologists have documented a number of examples of culture in a range of species: socially-learned practices such as complex nut-cracking by chimps in the Guinea forest; the tool-

---

[21] Even among social scientists definitions vary significantly; in 1952, A.L. Kroeber and C. Kluckholm published an article citing 164 different definitions of culture held by social scientists.

use of Sumatran orangutans; and self-medication in a variety of primates. Again, cultural practices are not limited to primates: Dorothy M. Fragaszy and Susan Perry's *The Biology of Traditions: Models and Evidence* published the findings of nearly a dozen separate studies of social learning and traditions among nonhuman creatures, from fish and dolphins to birds and rats.

As the title of Fragaszy and Perry's book suggests, not all biologists are comfortable with the use of the term "culture," despite the fact that the idea of nonhuman cultures has a great deal of support (primatologist William McGrew, a "pro-culturalist," has characterized this state of affairs as "the controversial, value-laden use of the 'c' word" [127]). In fact, in their introduction to *The Question of Animal Culture*, editors Laland and Bennett G. Galef refer to "the recent spate of articles in prominent scientific journals, newspapers, and news magazines that argue that differences in the behavioral repertoires of animals living in different locales provide evidence that they, like humans, are cultural beings" (1). While several researchers in the collection advocate the idea or actuality of nonhuman cultures without any or many qualifications, others do not, in part because of the interdisciplinary nature of this research:[22] there are varied methodologies, differing ideas of evidence, and basic definitional disagreements. One author's nonhuman "culture" is another's animal "tradition," "pre-cultural" practice, or social learning. Nevertheless, "There is nothing more circular than saying that we, humans, are the product of culture if culture is at the same time the product of us," de Waal and Kristin E. Bonnie argue in their chapter. "Natural selection has produced our species, including our cultural abilities, and hence these abilities fall squarely under biology. This inevitably raises the question whether natural selection may have produced similar abilities in more than one species" (19).

In the decades since Mito and Imanishi first discovered the cultural innovation of potato washing on Koshima Island, the macaques have shifted their practices by dipping their potatoes in the ocean, rather than freshwater. On a recent trip to the island, de Waal observed this first hand:

> Walking in shallow water, they would alternate dipping a potato in and chewing off a piece. They did not do much rubbing in the water, probably because these potatoes were prewashed: there was hardly any dirt to be removed. ... For this reason, Japanese scientists have changed their terminology ... Assuming that it is the salty taste of the water that the monkeys are after, they now speak of 'seasoning.' (204)

Not only have cultural practices now been documented, but even the evolution of such practices.

The most dramatic example of observed nonhuman cultural change doesn't focus on tool use or the transformation of food, but on large-scale social evolution

---

[22]   The emergent field of nonhuman social studies includes primatology, behavioral ecology, evolutionary biology, ethology, comparative psychology, and (to a lesser extent) anthropology.

(or, one might say revolution), documented by neurobiologist and primatologist Robert M. Sapolsky's ongoing research on an olive baboon troop in East Africa. His best-selling account of this work, *A Primate's Memoir*, chronicles the changing personalities and social structures of this group over two decades and its near-destruction by environmental poisoning; many of the male baboons, and some females, contracted bovine tuberculosis from infected meat and organs (tossed to them as scraps, and left in the garbage dump of a nearby tourist lodge). The most interesting finding of Sapolsky's research, however, follows the events described in this book.

Among primates, baboons are famously aggressive; as Sapolsky puts it, "they're no one's favorite species" (299). Sapolsky began studying the biology of stress through baboons in the first place because their societies, like the human society Sapolsky comes from (he is based in the American academy), are hierarchical and aggressive: "Basically, baboons [in the Serengeti] have about a half a dozen solid hours of sunlight a day to devote to being rotten to each other. Just like our society ... We live well enough to have the luxury to get ourselves sick with purely social, psychological stress" (15). However, in the years following the tuberculosis (TB) epidemic, which killed half of the males, including every alpha male, the culture of the troop changed radically. In the National Geographic documentary *Stress: Portrait of a Killer*, Sapolsky states that before the TB deaths, this troop was "your basic old baboon troop at the time, which means males were aggressive and society was highly stratified." Following the deaths, however,

> what you were left with was twice as many females as males, and the males who were remaining were, you know, just to use scientific jargon, they were good guys. They were not aggressive jerks. They were nice to the females. They were socially-affiliated. It completely transformed the atmosphere of the troop. And when new adolescent males joined the troop, they'd come in just as jerky as any adolescent males elsewhere on this planet, and it would take them about six months to learn we're not like that in this troop. We don't do stuff like that. We're not that aggressive. We spend more time grooming each other. Males are calmer with each other. You do not dump on a female if you are in a bad mood. And it takes these new guys about six months and they assimilate this style [of social life] and you have baboon culture. And this particular troop has a culture of very low levels of aggression and very high levels of social affiliation. And they're doing that 20 years later. (*Stress*)

Here, Sapolsky has found evidence that what he wrote in *A Primate's Memoir* about TB is also true of another biological product, of culture: "Biology in the lab is not biology in the wild" (287). It also suggests, as he has it in the documentary, "if they [these baboons] are able to in one generation transform what are supposed to be textbook social systems, sort of engraved in stone, we don't have an excuse when we say there are certain inevitabilities about human social systems."

Interestingly, even Sapolsky's groundbreaking work isn't free from science's traditional fear of anthropomorphism. "I was," he writes of his early research, "way too insecure in my science to publish technical papers using these names [for baboons]—everyone got a number then. But the rest of the time I *wallowed* in biblical names (14, italics mine). Sapolsky confronts this fear in the form of behavioral categories:

> Debates rage among animal behaviorists as to the appropriateness of using emotionally laden human terms to describe [nonhuman] animal behaviors. Debates as to whether ants really have "castes" and make "slaves," whether chimps carry out "wars." One group says the terms are a convenient shorthand for lengthier descriptions. One group says they are the same thing as human examples of these behaviors. Another group says they are very different, and that by saying that all sorts of species take "slaves," for example, one is subtly saying that it is a natural, widespread phenomenon. My bias is to agree somewhat with this final group. Nevertheless, Solomon did something that day that I think merits the emotionally-laden term that is typically used to describe a human pathology. Solomon chased Devorah, seized her near an acacia tree, and raped her. (24)

Throughout *A Primate's Memoir*, Sapolsky struggles with and, eventually, against this (as he writes) "bias" against anthropomorphism, concluding the book with a wish for the right Prayer for the Dead for the baboons he is unable to save (301).

The study of nonhuman cultures overlaps with the biological study of human cultures. The charge leveled at the former, anthropomorphism, is related to the charge of determinism leveled at sociobiology and its descendants.[23] In the first case, critics mischaracterize anthropomorphism as anthropocentrism, whereas de Waal distinguishes between "animalcentric anthropomorphism" and "anthropocentric anthropomorphism," saying, "The first [makes every effort to take] the animal's perspective, the second takes ours. It is a bit like people we all know, who buy us presents that they think *we* like versus people who buy us presents that *they* like. The latter have not yet reached a mature form of empathy, and perhaps never will" (77). He argues that if anthropomorphism is risky, "its opposite carries a risk too. To give it a name, I propose *anthropodenial* for the a priori rejection of shared characteristics between humans and animals when in fact they may exist" (de Waal 68–9, italics in original). While anthropodenial is still the default position of

---

[23]   In 1975, E.O. Wilson's *Sociobiology: the New Synthesis* applied Darwinian principles to human behavior. Wilson coined the term sociobiology, and from this discipline grew other evolutionary approaches to behavior: behavioral ecology, evolutionary psychology, and gene-culture co-evolution. For an explanation of the differences between these fields, see Laland and Brown's *Sense and Nonsense*. In *A Primate's Memoir*, Sapolsky argues, "Sociobiology is often faulted for the Machiavellian explanations it gives for some of the most disturbing of social behaviors. ... Less noticed is that it also generates just as valid (or invalid) explanations for some of the most selfless, altruistic, caring of behaviors and shows the circumstances under which those are highly rewarding behavioral strategies to follow" (101).

the sciences and the humanities, it may be a hard habit to maintain. As biologist Marc Bekoff writes of his colleagues,

> I know no practicing researcher who doesn't attribute emotions to their companion animals—who doesn't freely anthropomorphize—at home or at cocktail parties, regardless of what they do at work. (This anthropomorphizing is nothing to be ashamed of, by the way ... these scientists are simply doing what comes naturally. Anthropomorphizing is an evolved perceptual strategy; we've been shaped by natural selection to view animals in this way.) (10)

Just as our survival depends on the survival of a great many other creatures, it seems reasonable to assume this evolved capacity of anthropomorphism, and the biophilia it engenders, is necessary for human (and other animal) survival. Bekoff argues, "If we don't anthropomorphize, we lose important information. ... it is a necessity, but it also must be done carefully, consciously, empathetically, and biocentrically. We must make every attempt to maintain the animal's point of view" (124–5).

The second case is a variation on a theme if not a mirror image. If nonhuman cultural studies, or cultural biology, is mired in false, sentimental identifications, then the biological study of human behavior denies the unique significance of human thought and feeling by claiming a biological basis of culture—treating *us* like "mere" animals! In his 1996 retrospective *In Search of Nature*, E.O. Wilson defends the evolutionary study of human and other animal behavior from charges of determinism:

> Concern over the implications of sociobiology usually proves to be a simple misunderstanding about the nature of heredity. Let me try to set the matter straight as briefly but fairly as possible. *What the genes prescribe is not necessarily a particular behavior but the capacity to develop certain behaviors and, more than that, the tendency to develop then in various specified environments* ... It is this *pattern* of possibilities and probabilities that is inherited. (89–90, italics in original)

Laland and Brown concur: "When researchers talk about genetic influences on human behavior, they do not mean that the behavior is completely determined by genetic effects, that no other factors play a role in our development, or that a single gene is responsible for each behavior" (17). In fact, "developmental biologists are agreed that the very idea that an individual's behavior can be partitioned into nature and nurture components is nonsensical, as a multitude of interacting processes play a role in behavioral development" (18).

The notion of freedom fueling this charge of determinism is, at root, a notion of human supremacy only conceptually possible if the rest of the living world is determined. Both logic and daily experience suggest, however, that nothing is determined and, equally, nothing is "free." We fear biological determinism not only because of the use made of the idea in the past, but also because Western culture at large continues to attribute every action and desire of other animals to a reductive notion of their biology, summed up in the derogatory (and tautological) use of the term "instinct." It is a shorthand way of saying that *they* are machines, organic machines acting under the rubric of their design. This, of course, is no more true of "them" than of us. De Waal reminds us that if biology restricts our freedom, culture does so to the same extent. "And where do our cultural capacities come from?" he asks. "Don't they spring from the same source as the so-called instincts? ... Whereas we can fully expect that definitions of culture will keep changing to keep the apes [and other animals] out, the proposals heard thus far seem insufficient to do so" (236). Just how far some scientists will go to keep changing definitions of culture to keep the "riffraff" out is itself a question of culture.[24]

Perhaps those who expressed horror at Wilson's *Sociobiology*, scientists and scholars in the humanities alike,[25] did so not because, or simply because, they misunderstood the text (or, as Wilson has it, took the notion of heredity to be deterministic) but because of the most pervasive form of liberal humanism: anthropocentric rationalism. "To be anthropocentric," Wilson writes, "is to remain unaware of the limits of human nature, the significance of biological processes underlying human behavior, and the deeper meaning of long-term genetic evolution" (100).[26] Val Plumwood characterizes anthropocentric rationalism, this dominant form of reason, as "a doctrine about reason, its place at the apex of human life, and the practice of oppositional construction in relation to its 'others,' especially the body and nature, which are simultaneously relied upon but disavowed or taken for granted" (18). It is a doctrine of power for power, which erases the subjectivity of other beings, creating living "resources" available for consumption. While this functional "misunderstanding" of the world enables its domination, it also misunderstands the enabling conditions of human life, of embodiment and embeddedness, at our peril.

---

[24] In "An Ape Among Many: Animal Co-Authorship and Trans-species Epistemic Authority," Bradshaw writes that science has traditionally excluded nonhuman animals from the creation of knowledge and its application to their lives, even in environmental policy. There is new science, however, which includes other species in the project of human knowledge, challenging old epistemological assumptions about other animals. Bradshaw discusses languaged ape and human participatory action research (PAR) at the Great Ape Trust as one example of trans-species science, work that contradicts the idea that language and knowledge are properties unique to humans.

[25] For just one example, see Richard Levins and Richard Lewontin's otherwise intelligent *The Dialectical Biologist*, in which they dismiss Wilson as *wholly* reductive: "A recent avatar [of vulgar reductionism] is Wilson's (1978) claim that a scientific materialist explanation of human society and culture must be in terms of human genetic evolution and the Darwinian fitness of individuals" (134).

[26] Or, "culture is created and shaped by biological processes while the biological processes are simultaneously altered in response to cultural change" (Wilson 111).

"The question of the purpose of human life has been raised countless times; it has never yet received a satisfactory answer and perhaps does not admit of one," argues Freud. And yet,

> Nobody talks of the purpose of the lives of animals, unless, perhaps, it may be supposed to lie in being of service to man. But this view is not tenable either, for there are many animals of which man can make nothing, except to describe, classify, and study them; and innumerable species of animals have escaped even this use, since they existed and became extinct before man set eyes on them. (24)

Here Freud presents us with the story of the first human question (what is the purpose of human life?) as the very origin of culture. It only makes sense, then, that nobody talks of the purpose of the lives of animals. Our purpose, as our story goes—the story that seems the very foundation of Western culture—relies on their distinct lack of purpose. Whether the story is religious (God has made us in his image and our purpose is to please him) or teleological (we are the unique pinnacle of life on earth) or both does not make a substantive difference. In either case, this story is a defense-narrative, what Freud calls a *détour* en route to a mature, frank acceptance of human powerlessness and finitude: "If the believer finally sees himself obliged to speak of God's 'inscrutable decrees,' he is admitting that all that is left to him ... is an unconditional submission. And if he is prepared for that, he probably could have spared himself the *détour* he has made" (36). What Freud called the reality principle we might call the biological conditions of life: the fact that human beings are not deities, cannot master nature or control their fate, but are, in fact, animals that evolved and continue to evolve with other lifeforms. "This recognition," writes Freud, "does not [need to] have a paralyzing effect. On the contrary, it points the direction for our activity" (37).

The implications of cultural biology are far-reaching and radical: we do not have to look to the sky to see that we are not alone in the universe. In her field-making introduction to *The Ecocriticism Reader*, Cheryll Glotfelty writes, "In most literary theory 'the world' is synonymous with society—the social sphere. Ecocriticism expands the notion of 'the world' to include the entire ecosphere" (xix). We must take this formulation a step further: ecocriticism must not only expand our notion of "the world" but also of "the social." Although we are not the only species that use culture to alter our environment, we are at the moment the only one endangering the existence of a *great many others*. Despite Venter's pronouncement that his new bacterium "heralds the dawn of a new era in which new life is made to benefit humanity," the new era doesn't sound *so* very new; other lifeforms have long been made to benefit humanity. That is, made to benefit some of us, in the short term, with widespread suffering and the risk of more.

For political intervention in this historical, ecological crisis, in which a great many real beings suffer, we must change our conception of the human and the nonhuman, of animality itself. The postanthropological concepts and findings of cultural biology topple the humanist idea of culture perpetuated by various ecocriticisms and posthumanisms, and the humanities generally. The realization that the human animal is one of many lifeforms engaged in the interwoven (indeed, co-creating) processes of nature and culture (or naturecultures) is the first step toward a more materialist ecocultural analysis or posthumanist multiculturalism— toward concepts of subjectivity and knowledge, and knowledge itself, transformed by interconnected social and ecological worlds. It is a step toward a political sensibility in cultural theory and analysis attuned to anthropodenial as well as anthropomorphism, one willing to explore the messiness of needs and our responsibilities to similarity and to difference.

The Bildungsroman

Exploring the implications of cultural biology in the humanities requires a fresh look at the stories culture explicitly tells about itself. In the West, these stories fall into the genre of the Bildungsroman (literally, narrative of acculturation). The Bildungsroman, as Adorno argues for lyric poetry, is the most social when it seems the least so.[27] While explicitly the story of the origin and development of the individual, the Bildungsroman is also culture's own origin story, the humanist myth of its separation from and opposition to nature. In the examples of the genre read in this book, underneath the positioning of nature as the Other of culture lies the recognition of the deep interconnectedness of the cordoned-off worlds of our biology and all that we build, physically and conceptually. It is the recognition of the agency, and sometimes even subjectivity, of nonhuman nature.

Marc Redfield writes that "[m]onographs on the *Bildungsroman* appear regularly; without exception they possess introductory chapters in which the genre is characterized as a problem, but as one that the critic, for one reason or another, plans either to solve or ignore" (380). This book would seem little exception to Redfield's rule if it were another genre study, for the Bildungsroman is truly one of the most, if not the most, defined, redefined, reconstructed and contested subgenres in literary study. This is due, in part, to interesting (and much needed) revisionist work by feminist and postcolonial scholars and writers. But, only in part. The category of the Bildungsroman has from its inception presented, as Redfield suggests, "a problem" for *many* critics. While Lukács, as Susan Suleiman remarks, seemed to consider all novels variations of the genre (64),[28] James Hardin stresses the importance of the genre's Germanic roots, bemoaning the "careless" or "naïve" use of the term in Jerome Buckley's 1974 *Season of Youth* and similar works (and the more recent, "needlessly cavalier application" of the term in feminist scholarship, such as *The Voyage In*).[29] However, Hardin also notes that in the eighteenth century, Bildung meant formation, in a "broad, humanistic sense" (xi)—that is development, acculturation. It is this foundational sense of the term

---

[27] See "On Lyric Poetry and Society" in *Notes to Literature*, also referenced in Chapter 3. This way of reading "negatively" will become important from Chapter 3.

[28] See *The Theory of the Novel*.

[29] x, xviii.

that interests me, as the formation of the human itself. The Bildungsroman is humanism's story of becoming human as becoming part of culture, the humanist origin story of culture itself, of its self-creation out of nature.

And yet, this origin story, like others, contains (or fails to contain) its own contradictions. It is the premise of this book that the story of individual acculturation is always the story of culture—but it is the argument of this book that this is also the story of "nature," of our knowledge of human animality and nonhuman agency or subjectivity. The Bildung of the protagonist, or Bildungsheld, need not be positive nor successful, and the novel need not be Germanic nor even realist to be useful to this project. The tale, novels, and, more radically, narrative essay discussed in this book were chosen because they too seem interested in this fateful question of nature and culture, in various forms: ideas of the garden, materiality, the human body, animalization, and, of course, development. While these texts, save Jamaica Kincaid's *A Small Place*, are characterized by some critics as Bildungsromane, they might by others be considered counter or anti-Bildungsromane or, by others still (such as Hardin or Franco Moretti), quite outside the genre altogether.[30]

Perhaps the most pressing "problem" any text that creatively or critically engages the genre faces is the specter of Goethe's *Wilhelm Meisters Lehrjahre*—for many, *the* prototypical Bildungsroman. Susan Fraiman's *Unbecoming Women*[31] handles this problem rather well:

> The continual fetishizing of *Wilhelm Meister* as originary text, even by many revisionist critics, has not only defended as normative the single path of middle-class, male development described above, eclipsing all others; not only established a canon of overwhelmingly male-authored and male-centered texts. It has also to a larger degree fetishized Wilhelm himself. Indeed, there has been

---

[30] The term itself was coined in 1819 by Karl von Mortgenstern who, as Hardin suggests, "linked the word *Bildung* to the hero's development and experience, to his education, and to the *Bildung* of the reader" (xiii–xiv). This book follows the lead of both Mortgenstern and the feminist study Hardin attacks, *The Voyage In*, for its expansive use of the term. Yet, prior to any reformation, the genre is already bound up with so many others that locating its traditional borders is no small task—it overlaps with (or, for scholars such as Hardin, is contrasted with) the Entwicklungsroman (novel of development), Erziehungsroman (pedagogical novel), roman-a-thèse (didactic novel), conte philosophique (philosophical tale), Künstlerroman (novel of the artist's formation), and Bildungsgeschichte (novel of complex psychological development) to name a few. The Bildungsroman is, in turn, often subdivided, as in the classical, European (as opposed to German), English, female, parodistic, postcolonial, Caribbean, and postmodern Bildungsromane. See Melitta Gerhard's *Der deutsche Entwicklungsroman bis zu Goethes "Wilhelm Meister"* on the first two of these terms, and Jeffrey Sammons's "The Bildungsroman for Nonspecialists: An Attempt at Clarification," in Hardin's *Reflection and Action*, on the Bildungsmotiv and Bildungsgeschichte.

[31] *Unbecoming Women* examines conduct literature as well as fictional examples of the genre: "We should think of women's conduct books and novels as particularly contiguous and interpenetrating forms" (14).

a tendency to think of the tradition as a family not of texts but of personages:
Wilhelm and his kinsmen. The effect of this has been to define the genre in terms
of a single heroic figure and to privilege an approach that emphasizes character.
(9–10)[32]

In order to think *with* the Bildungsroman, to use it "as a conceptual tool" as *The
Voyage In* suggests,[33] one must avoid "renaturalizing the Germanic genre," as
Fraiman succinctly put it (144). And so it seems less useful to create the category
of "radical Bildungsroman" for these texts, radical though they are, than to
consider more broadly what they suggest about the complexity and contradictions
of humanism.

Given Glotfelty's foundational statement of the impetus for ecocritical work—
to move beyond the equation of the (human) social world with the world—this
genre seems ideal "terrain" for work on the nature of humanist culture. Moretti's
genre-study *The Way of the World* may as well have been titled *The Way of
Culture*, or the world *as* human culture. Its birth more or less congruent with that
of the Anthropocene, the Bildungsroman is, as Moretti famously articulates it, the
"'symbolic form' of modernity" (5); it makes modernity human (6),[34] as it makes the
human modern. It does so by being, of necessity, "*intrinsically contradictory*" (6):

> The success of the *Bildungsroman* suggests in fact that the truly central
> ideologies of our world are not in the least … intolerant, normative,
> monological, to be wholly submitted to or rejected. Quite the opposite: they
> are pliant and precarious, "weak" and "impure." When we remember that the
> *Bildungsroman*—the symbolic form that more than any other has portrayed
> and promoted modern socialization—is also the most contradictory of modern
> symbolic forms, we realize that in our world socialization itself consists first of
> all in the interiorization of contradiction. The next step, being not to "solve" the
> contradiction … but rather to learn to live with it, and even transform it into a
> tool for survival. (10)

I would take this formulation of the Bildungsroman as contradiction a step further
with Pierre Macherey's explanation of the contours of ideology in text from *A
Theory of Literary Production*:

---

[32]   Many critics have noted that not even *Wilhelm Meister* entirely fits the definition of
the genre it supposedly exemplifies.

[33]   "It has become a tradition among critics of the *Bildungsroman* to expand the concept
of the genre: first beyond the German prototypes, then beyond historical circumscription,
now beyond the notion of *Bildung* as male and beyond the form of the developmental
plot as a linear, foregrounded narrative structure. Our reformation participates in a critical
tradition by transforming a recognized historical and theoretical genre into a more flexible
category whose validity lies in its usefulness as a conceptual tool" (*The Voyage In* 13–14).

[34]   "Bourgeois freedom is peculiar in that is has generated the unceasing counter-melody
of the 'escape' from its harshness. The Frankfurt School … defined this ambivalence as a
chronological succession: first freedom—then the escape of which Erich Fromm wrote. …
[In the Bildungsroman] the dialectic of bourgeois freedom does not unfold as a succession
of 'first, then later'—but as the continual co-presence of the two opposing tensions" (66).

This conflict is not the sign of an imperfection; it reveals the inscription of an *otherness* in the work, through which it maintains a relationship with that which is not, that which happens at its margins. To explain the work is to show that, contrary to appearances, it is not independent but bears in its material substance the imprint of a determinate absence which is also the principle of its identity. The book is furrowed by the allusive presence of those other books against which it is elaborated; it circles about the presence of that which it cannot say, haunted by the absence of certain repressed words which make their return. The book is not the extension of meaning; it is generated from the incompatibility of several meanings, the strongest bond by which it is attached to reality, in a tense and ever renewed confrontation. (79–80)

One of the books—in fact, the chief "book"—against which all others are elaborated is "the book of nature."[35] When Moretti writes that the Bildungsroman creates the modern sense of "everyday life" as an "anthropocentric space" (12),[36] he approaches the edge of the idea that the Bildungsroman constructs, as suggested earlier, a vision of the world itself as (humanist) culture.

And so the narrative of the individual coming into culture is not only the story of the unquestioned "truth" of human separateness from and supremacy over the rest of nature; it is often (perhaps always) also the story of the struggle for and anxiety about this supremacy. As Redfield remarks, "the content of the genre is never simply a 'content,' but is always also '*Bildung*,' formation—the formation of the human as the producer of itself as form" (380).[37] In other words, it is, as I will argue, the formation of the human as the producer of itself *as culture*, the humanist equation of the human with culture (and culture as exclusively human).

---

[35] Michel Foucault writes in *The Order of Things*: "The great metaphor of the book that one opens, that one pores over and reads in order to know nature, is merely the reverse and visible side of another transference, and a much deeper one, which forces language to reside in the world, among the plants, the herbs, the stones, and the animals" (35).

[36] Moretti describes what he terms the classical Bildungsroman as conservative (59), concluding:

> "[m]eaning in the classical *Bildungsroman* has its price. And this price is freedom" (63), even as "[l]iberal thought itself coined a definition of freedom as 'freedom from'" (66).

[37] Following Redfield, Maria Helena Lima notes that the Bildungsroman is "recognizably one of the main carriers of humanist ideology" ("Imaginary Homelands ..." 859). Lima introduces the notion of "transculturation" to describe the way in which some, particularly postcolonial, cultures transform this "originary" genre to serve specific needs, "a corrective to anthropology's unidirectional 'acculturation.' Transculturation implies different cultural matrices impacting reciprocally on each other to produce a heterogeneous ensemble rather than a single culture" ("Decolonizing Genre ..." 433). Here Lima uses the term to describe the way in which postcolonial texts use the Bildungsroman to "write back" to the center; we will see an example of this in Chapter 5, with several of Kincaid's works. I am, however, more interested in the broader implications of "transculturation" in light of cultural biology—the idea that human and nonhuman cultures reciprocally impact each other, as "matrices"—or, as Darwin had it, a web of complex relations.

Chapters

The following chapters focus on four critically significant works that represent major literary movements of the modern era, Voltaire's *Candide*, Shelley's *Frankenstein*, Woolf's *Orlando*, and Kincaid's *A Small Place*. Candide enters culture by way of an Edenic garden, from which he is cast out for "original" sin with Cunégonde: "Their lips met, their eyes glowed, their knees trembled, their hands wandered. The Baron of Thunder-ten-tronck came around the partition and, seeing this cause and effect, drove Candide out of the castle" (Voltaire 43). In *Frankenstein*, a similar garden of domestic happiness precedes Victor Frankenstein's removal from Geneva to Ingolstadt, his intellectual isolation and unsuccessful struggle to return to community. All Frankenstein and his Monster can do is separate, die, or float off into "darkness and distance." Orlando too leaves a secure garden-estate, of which he is lord and master of land and beast, darling of Majesty and Empire: "His fathers had been noble since they had been at all. They came out of the northern mists wearing coronets on their heads ... From deed to deed, from glory to glory, from office to office he must go" (Woolf 11–12). Positioned rhetorically as a naïve tourist, "you," the reader of *A Small Place,* travel to Antigua only to learn that you always carry your home with you, that you are another colonist, that everything here (as at home) is on your terms, that as you stop to admire beauty, which to you has no history and which you cannot understand, you are an "ugly" thing. Driven from the rationalizations of imperialism and global capitalism, the reader is cast into the Other's experience of the dominated world. These narratives of arrival and development in humanist culture critique the metanarrative of Western culture's origin and value.

Chapter 2 demonstrates the way in which an ecocritical approach to Voltaire's *Candide* complicates Marxist materialism's (specifically Horkheimer and Adorno's in *Dialectic of Enlightenment*) approach to Enlightenment texts while enabling the spirit of its critique. Through a close reading of Mary Shelley's *Frankenstein* (alongside Percy Bysshe Shelley's "Mont Blanc"), Chapter 3 does the reverse, examining the way in which a dialectical approach to culture both complicates and completes ecocritical readings of Romantic literature. These chapters focus on texts that are themselves philosophically significant and representative of their intellectual age—*Candide* is, in many respects, the quintessential Enlightenment narrative, just as "Mont Blanc" and *Frankenstein* are exemplary texts of literary Romanticism.

Contrary to generalizations of the Enlightenment, a clear link exists between Voltaire's critique of rationalism and ecological critiques of rationalism. *Candide* repeatedly demonstrates that we are neither the masters of the earth nor of our fate—and yet Voltaire ultimately casts his assertion of the limits of reason, knowledge, and human power in positive terms. Candide's education concludes with the first exercise of his newly gained wisdom: he insists, "*mais il faut cultiver notre jardin.*"[38] His injunction responds not only to Pangloss's blind optimism

---

[38]   Translation: "but we must cultivate our garden."

and Martin's visionless pessimism, and to systematic theorizing altogether, but also to the natural and social crises witnessed on his voyage. In the context of a tale that insists on the inescapability of material reality, and in the historical context of Voltaire's own commitment to cultivation, it makes little sense to read Candide's dictum as mere metaphor or transcendental ideal; whatever else it may signify, the garden must also signify a real, material garden. Although *Dialectic of Enlightenment* identifies and traces the nexus of rationalism and power that dominates nature from Bacon to the culture industry (and even suggests that this domination of nature may come to mean the end of culture), the text follows the movement of history *solely as human history* when, in fact, human history is always already inextricably interwoven with and embedded in nonhuman histories.

Chapter 3 examines *Frankenstein* not only as a critique of scientific technology and imperialism, but also as an articulation of anxiety, similar to Percy Bysshe Shelley's "Mont Blanc" (originally published in Mary and Percy Shelley's joint record of their trip to the Alps during which the first draft of *Frankenstein* was written, *History of a Six Weeks' Tour*). The narrative structure of *Frankenstein* embeds human activity in, quite literally, a sea of connections. The novel is surrounded by the Arctic Ocean and, within each concentric narration, a body of water serves as the background for the novel's most dramatic action; either Lake Leman, the North Sea, or the Arctic surround the meetings between Victor and the monster. Here, expanses of water function as sublime landscapes in their own right, as unfathomable immensities and/or impenetrable depths. A dialectical approach to ecocritical concerns in the text reveals an important connection between these natural bodies of water and the produced body of the monster. Victor's drive to transcend human nature and culture, to transcend the limitations of the human body and human knowledge is akin to Robert Walton's drive to transverse the Arctic Ocean. In both cases, materiality becomes an obstacle to overcome, rather than the fabric of existence. The constant narrative proximity to bodies of water and its parallel in the body of the monster serve to remind us of our bodies, and the human place in the material weight of the world.

Of course, Shelley's novel is not about human bodies alone. Our experiences of the world (of the nonhuman and, indeed, the human) may be negatively inscribed, or unconscious, in culture. "Positively" and "negatively" inscribed responses to nonhuman nature, somewhat analogous to photographic images recorded as negatives and developed as positives, coexist in the text simultaneously. Reading "negatively," then, suggests attending to both aspects of a text; it is an imperfect analogy, but one not wholly unlike Marx's *camera obscura* in *The German Ideology*.[39] This critical position and methodology suggests that the nonhuman not only encompasses and impacts human culture, in ways that we can and cannot see,

---

[39] "Consciousness can never be anything other than the conscious being, and the being of men in their real life process. If, in all ideology, men and their relationships appear upside down, as in a *camera obscura*, then this phenomenon stems just as much from their historical life process as the inversion of objects on the retina stems from the processes of direct physical life" (*Marx: Early Political Writings* 124–5).

but that it also might serve as an intervention in human culture, in ways that we both can and cannot understand. There is much negatively inscribed in *Frankenstein*, what the text means but cannot always understand or say. Sometimes, as Macherey suggests, "the work cannot speak of the more or less complex opposition which structures it; though it is its expression and embodiment. In its every particle, the work *manifests*, uncovers, what it cannot say" (84). Inversely, the monster or the monstrous is that which speaks but isn't allowed to "mean." Monstrosity is constituted by meaningful acts of anything, anyone *not human*; the monster, the ultimate Other in Western culture, serves to deny the existence of this worldly nonhuman agency, to relegate it to the shadows and the fantastic. And yet, while the monster signifies many things (including the unsignifiable itself), he is also the voice of humanist culture betraying its fear of nonhuman agency.

Chapter 4 examines a "high" Modernist text to investigate Modernism's critique of modernity. *Orlando*, in part a mock-biography of Vita Sackville-West, journeys beyond the male literary canon and Carlyle's formulation that "history is the story of great men" by challenging Freud's claim that "anatomy is destiny." As a treatise against patriarchal history and biological determinism, *Orlando* offers an alternative vision of biography and biology, an ecological vision of human consciousness and material interconnectedness. As an experimental modern novel, it is an apt case study for an ecocultural approach because it represents a movement that has traditionally been difficult for both ecocriticism and Marxist materialism (cultural materialism and Frankfurt Theory) to theorize. *Orlando* presents a unique opportunity to consider a text that relentlessly moves a character through time and space but with the result, I argue, of making environment (human and nonhuman) central to the text. Alongside aesthetic differences, one of the characteristics used to define Modernism has been its international cosmopolitanism or, in contrast to Romanticism, its seeming placelessness. The chapter argues that the supposed "placelessness" of Modernism is itself a placeholder for anxiety about modernity, in all its fast-paced fragmentation and concentrated urban confusion. An ecocultural materialist examination of *Orlando* suggests that this quality of placelessness expresses modernity's negation of place as a locus of meaning and its broader anxiety about the agency of the more-than-human world.

Through Orlando's travel to the East,[40] change of gender, and supernatural experience of history, the novel takes as its subject the false duality of nature and culture in Western thought and representation. Read alongside *To the Lighthouse* and another radical Bildungsroman, a biography of Elizabeth Barrett Browning's spaniel *Flush, Orlando* is especially suited to a rethinking of prevailing critical approaches to Western culture.

Chapter 5 focuses on Jamaica Kincaid's *A Small Place*, a narrative of the colonial history and postcolonial life of Antigua. In the context of *Annie John* and *Lucy*, her traditional Bildungsromane, I argue that *A Small Place* may be read as

---

[40]   I will refer to İstanbul throughout the text as Constantinople, as Voltaire does in *Candide* and Woolf in *Orlando*. İstanbul became the official name of the city in 1930.

a new form of the genre in two connected senses. First, positioned rhetorically as a tourist in Antigua, the reader "comes into culture" as a naïve consumer and leaves an interpreter. "You" (the reader) are lead to recognize false histories and ideologies of Western culture, constructions masquerading as "nature," through the past and present of Antigua and its inhabitants. This lesson, echoing the didactic roots of the genre, juxtaposes the experience of "you" the tourist with the experiences of native Antiguans. Second, while a traditional Bildungsroman often depicts an individual leaving home to travel or traveling to learn or work, *A Small Place* uses the conceit of the reader as tourist to dramatize the fate of someone else's home, of an entire place and people. It uses travel to tell a story about those that have been invaded and relocated, those that are too poor to travel and "too poor to live properly in the place they live" (19). In this way, the narrative links the genre's trope of travel and individual development (examined in eighteenth-, nineteenth-, and early twentieth-century Bildungsromane in Chapters 2 through 4) to the current ecological and socio-economic realities of tourism and the maldevelopment of the postcolonial world. Read as the Bildungsroman of the reader-as-tourist, and the Bildungsroman of an actual place and people forced into the global economy, Kincaid's narrative echoes and pushes the traditional boundaries of the genre.

Chapter 5 concludes with an examination of Kincaid's 2005 travel narrative *Among Flowers* as a companion to *A Small Place*, but one in which the political anger of the latter is nowhere to be found. The narrative embodiment that rescues the violence of *A Small Place* from callous abstraction (the voice of the perpetrator of violence speaking as a human voice, in history) is entangled with depoliticizing embodiments of the tourist/colonizer in *My Garden (Book):* and *Among Flowers*, from Kincaid's own anxiety of human embodiment to her "animalizing" view of other native peoples, nonhuman animals, and ecosystems. While Kincaid's work generally critiques Enlightenment thought and ideas, it is also entangled with its most foundational error.

Chapter 6 concludes the book with a consideration of the issues of dehumanization raised in Chapter 5 through an examination of the connection between dehumanization, animalization, and ideas of empathy. Both the philosophy of the biology of dehumanization and the mainstream science of empathy participate in an unexamined discourse of animality. I argue that our ideas about empathy are bound up with this discourse and vice-versa, a connection that reveals a cultural fantasy of detachment. In the West, this fantasy of detachment is in fact the very definition of the human (as free will, the freedom to surpass nature, the freedom to master) and the beast (as unrestrained drives, as instinct without empathy). Positioning other animals as agents without empathy allows us to imagine them outside the ethical community, depriving them of our sympathy, keeping them outside the polis, denying them politics. The humanist discourse of culture, then, is not simply, or only, about the uniqueness of human reason, but also about the uniqueness of human social feeling. Political relations, the hope of sharing together the polis we live in, requires more than discourses of interest and rights; it requires a challenge to the fantasy of the human itself, the discourse of animality that fuels the animalization of human and nonhuman beings.

The last chapter ends with a return to the idea of origins. Eden is, for the Western tradition, the "original" origin story. For the Bildungsroman, it is not only the garden to which all others must refer, the origin story of human supremacy over the rest of nature (as Lynn White Jr.[41] and so many others have noted), it is also the Ur-plot of human "development," the expulsion from childhood into adulthood, from nature into culture. It is, in this way, also the framework for the humanist origin story of culture—of its creation of itself from what Donna Haraway calls "the soil of nature." Revisiting Kincaid's *Among Flowers* and *My Garden (Book):* (examined in Chapter 5) permits a consideration of "gardening" in the broadest sense in the previous texts. Contrasting Jacques Derrida's refutation of "the animal" with Slavoj Žižek's argument against nature allows a critique of the latter in the context of the implications of the biological idea of culture.

Again, although we are *not* the only species that uses culture to alter our environment, we are at the moment the only one endangering the existence of almost everyone else. For true intervention in this historical, ecological crisis, we need that which Kafka attributed to the best literature, "an axe to break the frozen sea within us"—a critical edge to change our conception of the human and the nonhuman. For the humanities, this means a new concept of culture.

The Cultural Unconscious

> Whether we grant animals culture is ultimately a human cultural question.
> —Frans de Waal, *The Ape and Sushi Master*

Ecocultural materialism diverges most significantly from Marxist materialism by rejecting the notion that "nature" is simply mediated by "culture." In keeping with cultural biology, this old idea of mediation is replaced with a greater skepticism about human uniqueness, an openness to the world as a true "web of complex relations," in which "nature" and "culture" mediate each other as complexes of co-creating nonhuman and human naturecultures. In this sense, the immanent world is perhaps "immediate" after all—that is, mediation is itself the fabric of the world. While Horkheimer and Adorno repeatedly stress that any appearance of immediacy is false, and assume other animals have no "selves," only compulsions (*Dialectic* 205), they also suggest that the process and perception of mediation has the potential to heal the human dominated world:

> Only mediation, in which the insignificant sense datum raises thought to the fullest productivity of which it is capable, and in which, conversely, thought gives itself up without reservation to the overwhelming impression—only mediation can overcome the isolation which ails the whole of nature. Neither the certainty untroubled by thought, nor the preconceptual unity of perception

---

[41] See "The Historical Roots of Our Ecological Crisis."

and object, but only their self-reflective antithesis contains the possibility of reconciliation. (*Dialectic* 156)

Without this dialectical reflection, reason turns back upon itself, and "humanity's sharpened intellectual apparatus is turned once more against humanity, regressing to the blind instrument of hostility it was in animal prehistory, and as which, for the species, it has never ceased to operate in relation to the rest of nature" (*Dialectic* 156). Horkheimer and Adorno warn that, on our present course, "the human species will tear itself to pieces or it will take all the earth's flora and fauna down with it" (*Dialectic* 186).

To argue that culture mediates nature (as a one-way process) or to assert that everything is nature (in an undifferentiated way) erases true political relations between human and other animal beings. The human experience of the world is mediated by our cultures, but these cultures are mediated by what we call nature: overlapping sets of human and nonhuman cultures amid wind, rain, and rock. The social networks and practices of myriad species transform the material conditions of life for themselves and the other inhabitants of the planet every day. Not only is everything and everyone in *some sense* interconnected, we all materially and culturally impact each other.

This book, however, is not a study of animal cultures. It is an attempt at an analysis of human culture informed by the existence of nonhuman cultures. It is an effort to take biology and ecology seriously, to integrate key notions of culture from the sciences and humanities to examine the stories humanist culture tells about itself. As Adorno teaches, art may reveal what ideology hides. In this case study, the Bildungsroman reveals an awareness of nature's agency, and human and nonhuman similarity. An anxiety about other animal subjectivity is underneath the resentment of the study of humans as animals and the conditioned dismissal of nonhuman subjectivities and cultures. It is an anxiety about the lie of human superiority and the fragility of power. It is the chief target of ecocultural analysis and the reason for this book.

Chapter 2
Candide and the *Dialectic of Enlightenment*

I write in order to act.
—Voltaire, Letter to Vernes, 15 April 1767

Voltaire makes the very act of living lucidly a form of active revolt.
—Patrick Henry, *Voltaire and Camus:*
the Limits of Reason and the Awareness of Absurdity

In "Why Still Philosophy," Adorno maintained that philosophy "should not with foolish arrogance set about collecting information and then take a position; rather it must unrestrictedly, without recourse to some mental refuge, experience." Philosophy must experience the world fully because history "promises no salvation and offers the possibility of hope only to the concept whose movement follows history's path to the very extreme" (*Critical Models* 17). While Horkheimer and Adorno's *Dialectic of Enlightenment* traces the nexus of rationalism and power that dominates nature from Bacon to the culture industry, and suggests that this domination may come to mean the end of history, the text follows history as a *human* path through nature when, in fact, it is always inextricably interwoven with and embedded in nonhuman histories.

Horkheimer and Adorno's critique of Western culture foregrounds the human domination of nonhuman nature as the logic of all domination. However, *Dialectic*'s conclusion about human (specifically, humanist) culture leaves little room for hope. In this reified[1] existence, the culture industry seems to co-opt or commodify every possible intervention into the hegemony of global capitalism; while *Dialectic* demonstrates the dire need for intervention, it also seems to preclude its possibility. Intervention will not, Horkheimer and Adorno argue, come from Marx's vision of the proletariat acting as "the motor of history" or from anything "inside" human culture. However, our lived experience of the world suggests cultural change may still come from what seems "outside" culture; nonhuman natures intervene in our lives daily, despite our attempts to contain them.

On November 1, 1755, a tremendous earthquake and ensuing tidal wave and fire killed between thirty and forty thousand people in Lisbon and the surrounding

---

[1] The *Oxford English Dictionary* defines reify as follows: "To make (something abstract) more concrete or real; to regard or treat (an idea, concept, etc.) as if having material existence." I use the term in this way, and also to describe a corresponding but seemingly opposite process, in which real, material entities are made less real; that is, the objectification of any dynamic or living entity (its treatment or conceptualization as a static, dead object).

towns (Havens 77). On November 24, 1755, François-Marie Arouet de Voltaire wrote to his friend Jean-Robert Tronchin,

> People will find it awkward indeed to explain how the laws of motion bring about such frightful disasters in "the best of all possible worlds." A hundred thousand ants, our neighbors, crushed in a second in our ant-hill, and half of them undoubtedly perishing in inexpressible anguish in debris from which it was impossible to extricate them: families all over Europe ruined, the fortunes of a hundred merchants from your homeland swallowed up in the ruins of Lisbon. What a sad game of chance is the game of human life! ... I hope that at least the Reverend Fathers, the Inquisitors, have been crushed like the others. That ought to teach men not to persecute men, for while a few holy scoundrels burn a few fanatics, the earth swallows up them all. (*Correspondence and Related Documents* 100: 401–2)

Shortly thereafter, Voltaire wrote the widely read *Poème sur le désastre de Lisbonne*, which expresses much the same idea as his letter to Tronchin.[2] Within three years of the poem's publication, Voltaire wrote and published his most influential critique of rationalism, *Candide: or, Optimism*, in 1759. An expression of Voltaire's vision of the world as an anthill on which the powerful oppress the powerless, *Candide* exemplifies the relevance of eighteenth-century thought to contemporary discourses of engagement. As a response to the intervention of nonhuman nature in human culture, *Candide* reflects the hope contained in the realization that human cultures are inscribed within a larger, more complex world. *Candide* exhibits several of the traits of the "traditional" or Germanic Bildungsroman, particularly the depiction of the development of an individual through travel.[3] As a catalogue of the horrors of the modern world, *Candide*— perhaps more than any of the other texts examined in this book—lives up to Moretti's articulation of the Bildungsroman as the "'symbolic form' of modernity" (5). Read from an ecocultural perspective, this philosophical Bildungsroman

---

[2]   An early draft of the poem had been finished by December 7th (Kendrick 119).

[3]   Giovanna Summerfield and Lisa Downward's *New Perspectives on the European Bildungsroman* also discusses *Candide* as an example of the genre, despite the general tendency to refer to it only as a conte philosophique; as noted in the Introduction of this book, the Bildungsroman and conte philosophique are overlapping genres. Many eighteenth-century scholars claim Voltaire invented the conte; however, one must consider the case of *Gulliver's Travels*, which strengthens the connection between the two genres. In 1726 Jonathan Swift traveled to England with the manuscript and published it anonymously that same year (*Travels* was then translated into French, German, and Dutch in 1727). In March 1728, decades before he penned *Candide*, Voltaire wrote to Swift (whom he had known in London): "the more I read your works, the more I am ashamed of mine." It is likely that Voltaire read Swift in English rather than in translation; as he wrote to Swift in December 1727, "pray forgive an admirer of you, who owes to yr writings the love he bears to yr language ..." *Gulliver's Travels* and *Candide* have a great deal in common: both are travel narratives of development, both satirize specific and general social problems, and both engage philosophical dilemmas.

suggests the limitations of *Dialectic*'s conceptions of the Enlightenment and the subject with a model, albeit a modest one, for interaction with the world outside of rationalism's logic of domination.[4]

Enlightenment Reconsidered

Throughout his lifetime, Voltaire struggled with many of the questions that ecological thinkers continue to ask in the twenty-first century: What are the limits of human reason and knowledge? What is the "good" life and how do we achieve it in an uncertain world? Voltaire's Bildungsroman demonstrates the absurdity of attempting to answer such questions through metaphysics and other forms of systematic rationalism. With *Candide*, Voltaire attacks Leibnizian optimism and seventeenth-century systematic rationalism in general, dramatizing a primary concern of the Enlightenment. As Ira O. Wade proposes in *The Intellectual Origins of the French Enlightenment*,

> [T]he central problem of the Enlightenment would be to grasp the way in which rationalism, having set out to become aware of its possibilities, has encountered hidden forces within itself which have made it conscious of its impossibilities. Simply stated, it is the story [of] how the human mind came to know and to turn into realities its inner powers, but how, in doing so it discovered not only their ultimate unreality, but their uselessness in achieving human satisfactions. (16)

Candide repeatedly demonstrates that we are neither the masters of the earth nor of our fate—and yet Voltaire, I argue, ultimately casts his assertion of the limits of reason, knowledge, and human power in positive terms. Candide's education concludes with the first exercise of his newly gained wisdom: he insists, *"mais il faut cultiver notre jardin."* His injunction responds not only to Pangloss's blind optimism and Martin's visionless pessimism, and to systematic theorizing altogether, but also to the natural and social crises witnessed on his voyage.

A long history of inquiry into the meaning of the text's conclusion has produced strikingly varied interpretations. Many critics view the instruction to "cultivate our garden" as a positive call for community action despite uncertainty or evil,

---

[4]   *Candide* seems an ideal case-study for a reexamination of the formulation of the dialectic of Enlightenment for several reasons. Aspects of *Candide*, the title character's wanderings and the encyclopedia of suffering depicted in the text, formally resemble Voltaire's other key work, noted by Henry Pickford in his introduction to Adorno's *Critical Models*: "When Adorno upholds that 'the element of the homme de lettres, disparaged by a petty bourgeois scientific ethos, is indispensable to thought,' he is invoking a German tradition in neo-Marxist essayism that effloresced in the Weimar Republic but that reaches back via Nietzsche to the figure of the French Enlightenment moralist and the discursive form of the nonsystematic critique, as in Voltaire's *Philosophical Dictionary*; in an analogy he repeats in his introduction to *Catchwords*, Adorno says of negative dialectics that 'thinking as an encyclopedia, rationally organized and nonetheless discontinuous, unsystematic, loose, expresses the self-critical spirit of reason'" (x).

while others argue that Candide's final statement reflects pessimism, resignation, or ambiguity.[5] Invariably, however, both eighteenth-century scholars and theorists read Candide's injunction as metaphor. Whatever else it may signify, the garden, I contend, must also signify a real, material garden—and, as we shall see, Voltaire's actual experience with gardens supports this claim. In the context of a tale that insists on the inescapability of material reality, it makes no sense at all to read Candide's dictum as mere metaphor or transcendental ideal. *Candide*, as Daniel Gordon points out, is not only *not* a metanarrative, it flatly rejects the metanarrative most commonly attributed to the Enlightenment, that of the progress and unification of human knowledge and the resulting mastery of nature (Gordon 202).[6]

Karlis Racevskis presents a strong case for *Candide*'s critical relevance, drawing a number of important parallels between the mature Candide and the contemporary critic. Both eschew abstractions in favor of a practical reason; both realize that solutions to social problems must be grassroots and communal, not dictated by intellectuals or experts; and both share the knowledge that work isn't protected by transcendental reason, destiny, or "a nature that will ensure humanity's survival in the face of the greed, cruelty, and stupidity of humans" (85–6).[7] This view of *Candide* supports an argument popular among certain eighteenth-century scholars, the idea that the dialectic of enlightenment should be understood "as a process internal to the Enlightenment—a process in which a certain degree of historical optimism immediately produced doubts about the completeness of the society desired" (Gordon 204). Here we have, almost precisely, the critical program of *Candide*. Finally eschewing abstract philosophies, Candide grounds

---

[5]   For example, Jerry L. Curtis, D. Langdon, William F. Bottiglia, and Patrick Henry view the text's final statement as a positive assertion; John Pappas and Giovanni Gullace read the same statement as negative or resigned, while other critics, such as Ira O. Wade and Geoffrey Murray, claim the ending to be ambivalent or ambiguous (Waldinger 19).

[6]   *Candide* presents a picture of a world that does not remain static as we try to master it; Candide's search for Cunégonde demonstrates that "what we gain is never what we anticipated, and this holds true for philosophy as well as for love" (Gordon 202).

[7]   Karlis Racevskis's *Postmodernism and the Search for Enlightenment* begins the ambitious task of formulating a critical middle ground between postmodernism and the Enlightenment with an affirmation of the value of Horkheimer and Adorno's critique in the context of other contemporary critiques of reason: "[T]heir argument proposed that 'reason did not necessarily produce rationality; on the contrary, the very project of rational control, while other social relations remained unchanged, produced unreason.' As a result of this basic flaw in the Enlightenment project, the alleged emancipation of 'Man' was also accompanied by the development of a 'manipulative, instrumental reason.' The suspicion that reason may have served as a convenient cover for processes and programs that had their own reasons and interests is also characteristic of the French structuralist and poststructuralist writers. The critique of reason becomes, for them, a component of a more general attack against the philosophy of humanism—a form of rationality whose elaboration is seen to accompany the development of certain strategies of political power in the nineteenth century" (3–4). This critique of the Enlightenment (or, as Gordon demonstrates, of seventeenth-century philosophy and science, and its eighteenth-century proponents) was inaugurated by Voltaire's philosophical skepticism.

himself materially through a commitment to shared meaning making, shared epistemological beliefs, fulfilling needs through communal work and life.

Patrick Henry writes in *Voltaire and Camus: the Limits of Reason and the Awareness of Absurdity*, "When not linked with experience, for example, we have seen that Voltaire not only considers reason an ineffectual instrument … but also a potentially dangerous one … [Voltaire] also insisted upon the inability of reason to systematize reality" (50). It is this very form of reason, and the resulting abstraction of ethics, that allows for the new world economy, which, as Candide learns, brings sugar to Europe but makes slaves in the process. In chapter nineteen, Candide encounters a horribly mutilated African slave in Surinam,

> "Oh my God!" Candide said to him in Dutch. "What are you doing in this horrible condition, my friend?"
>
> "I'm waiting for my master, Mr. Vanderdendur, the famous merchant," replied the Negro.
>
> "Was it Mr. Vanderdendur," asked Candide, "who treated you this way?"
>
> "Yes, sir," said the Negro, "it's normal. They give us one pair of linen trunks twice a year as our only clothing. When we work in the sugar mills and catch our fingers in the grinder, they cut off our hand. When we try to escape, they cut off our leg. I've had both punishments. It is at this price that you eat sugar in Europe." (*Candide* 83)

Perhaps the most horrifying part of the slave's condition is his own appraisal of it, summed up in the two words, "it's normal."

Voltaire's insistence on "pragmatic moral eclecticism" over lifeless abstractions—the abstraction of ethics that condones and creates violence and oppression—demonstrates, as Henry argues, that "Voltaire wanted to establish a way of life, not a body of thought. His main concerns were practical, not speculative" (89–91). *Candide* satirizes the rationalist philosophies of the previous century and suggests that the communal practice of a materialist realism is the best road to social harmony. The fact that Voltaire chose gardening, in his Bildungsroman and in his life, as the embodiment and practice of this realism is indeed significant.

In many ways, the Enlightenment and ecological critiques of rationalism are similar. Compare, for instance, their views on Cartesian dualism. Voltaire staunchly rejected metaphysical epistemology, and two of Descartes's most basic concepts in particular, that of animals as *animaux-machines* and humans as *res cogitans*. As Henry writes, "Voltaire not only considered this opinion to be empirically untenable, but also perceived that it created a cleavage in the zoological chain of being and was essentially an offshoot of the old anthropocentric vision of the world" (35).[8] While Voltaire critiques anthropocentrism, Val Plumwood laments its mirror image, the debasement of nonhuman nature:

---

[8] Henry arrives at this conclusion based on a great many sources, including *Micromégas, Lettres philosophiques*, and several volumes of correspondence.

> Cartesian thought has stripped nature of the intentional and mindlike qualities which make an ethical response possible. Once nature is reconceived as capable of agency and intentionality, and human identity is reconceived in less polarized and disembodied ways, the great gulf which Cartesian thought established between the conscious, mindful human sphere and the mindless, clockwork natural one disappears. (*Feminism* 5)

The ecological critique of rationalism, consciously or not, operates within a tradition of materialist skepticism that owes much to Voltaire's thought.

Here is another, particularly lucid example. In *Feminism and Ecology*, Mary Mellor argues that historical materiality, the embodied and ecological embeddedness of human existence, may serve as the basis for an "immanent realism." For ecofeminist Mary O'Brien, and for Mellor, the real work isn't "some abstract humanism, but ... the regeneration of and reintegration of historical and natural worlds" (O'Brien, qtd. in Mellor 91).[9] This is, fundamentally, an argument against Western, dualistic epistemologies that cast knowledge as transcendent, over and against the reality of the larger world; it is an argument for an "immanent realism ... revealed through patterns of subjugation and the perspectives they generate within the human community, and through an awareness of the interrelatedness of humanity and nature in ecological processes" (Mellor 111). For ecological thinkers, human existence is "framed in radical uncertainty about material conditions in their widest and deepest sense" (Mellor 148). This radical skepticism forms the epistemological and ontological core of materialist ecological thought. A material, subjective ontology is not, in this framework, a metanarrative, but rather a "grounding awareness" (Mellor 185). And such awareness isn't deterministic but processual: Mellor writes, "Neither humanity nor 'nature' are determinant; what is inescapable are the consequences of the dynamics between them" (13).[10]

---

[9]   I use the phrase "real work" with reference to Gary Snyder's concept of "what needs to be done" to rescue and sustain local and global ecological health (cf. *The Real Work: Interviews and Talks 1964–1979*, *The Practice of the Wild*, and *A Place in Space*).

[10]   Of course, Voltaire's view of the proper use of reason can differ sharply from that of materialist ecofeminism (and critical posthumanism), as in the concept of rights. Both Mellor and Plumwood, for example, find the liberal discourse of rights inescapably problematic for theoretical and historical reasons that have been discussed at great length elsewhere (for example, see Alison M. Jaggar's chapter on liberal theory and feminism in *Feminist Politics and Human Nature*, Maria Mies's and Vandana Shiva's chapters on liberalization in *Ecofeminism*, and Cary Wolfe's *Animal Rites*—discussed later in this book). Here, Plumwood critiques ethical universalization and its discourse of rights that still dominate the field of environmental ethics: "Universalization ... is the moral complement to the account of the self as "disembodied and disembedded," as the autonomous self of liberal theory, the rational egoist of market theory ... A more promising approach for an ethics of nature ... would be to remove rights from the center of the moral stage and pay more attention to some other, less dualistic, moral concepts such as respect, sympathy, care, concern, compassion, gratitude, friendship, and responsibility ... They are moral "feelings"

The Materiality of the Garden

Voltaire's *contes* (which include *Zadig* and *Micromégas*) satirize philosophical ideas and social institutions. Most famously, *Candide* attacks a particular concept of optimism, akin to Alexander Pope's claim from *Essay on Man*, "Whatever is, is right" or, as Pangloss has it, the belief that we live in "the best of all possible worlds." This is only one of a long list of targets, which includes Catholic clergy, provincial nobility, and the pandemic institutions of slavery, prostitution, and war.

As the pedagogical fulcrum of the tale, Candide is the only character who may be said to develop; his actions anchor every scene and chapter, and he is the primary audience for every narration of physical suffering and catastrophe. These narratives, combined with the suffering and catastrophe Candide himself witnesses on his journey—storm, shipwreck, earthquake, war, and disease—layer to achieve the tale's philosophical theme: the inescapability of material reality and, therefore, the benefit of cultivating one's garden.

Many male characters relate harrowing tales of woe, but the most significant of all is told by an old woman. After Cunégonde recounts her misfortunes to Candide, her servant boasts, "Alas! You have never seen misfortunes like mine" (*Candide* 60). In the course of chapters eleven and twelve, the old woman recounts her kidnapping, bondage, repeated rape, mutilation, and bout of plague. Her narrative stands out as particularly awful, and she is the only character in the tale, aside from Candide, who achieves any wisdom. The narrator repeatedly describes her as prudent and, tellingly, it is she that Candide consults for advice in Pangloss's absence (*Candide* 57–8, 67). In other words, this old woman who speaks "from age and experience" serves as a foil to Pangloss's and Martin's abstractions (*Candide* 67).

On the surface it might seem troubling, from any feminist perspective, that Voltaire chooses male characters (and only male characters) to represent philosophical concepts and an old woman to represent experience as a lifetime of suffering. It certainly would not be the first time "woman" has been made to bear the weight of the world's—particularly men's—materiality. From Eve's "fall" in the Garden of Eden to the eighteenth-century salon quip that women are but large children (occasionally attributed to Rousseau), Western culture has long associated women with materiality as opposed to intellect, particularly the "sins" and suffering of the flesh. However, nearly everyone in the tale, male and female, suffers physically—both Cunégonde and her brother are stabbed and raped, Pangloss is whipped and hanged. In the final chapter, the old woman recounts their collective suffering in one stroke:

---

but they involve reason, behavior, and emotion in ways that do not seem separable" ("Nature" 6, 8–9). Plumwood argues that an "ethic of care and responsibility" extends far more easily to nonhuman nature than the discourse of rights, and provides a better basis for non-instrumental relations with the world (9).

> I'd like to know which is worse—to be raped a hundred times by Negro pirates,
> to have a buttock cut off, to run the gauntlet among the Bulgars, to be whipped
> and hanged in an auto-da-fé, to be dissected, to row on the galleys, in short, to
> experience every misfortune we have known—or to stay here without anything
> to do? (*Candide* 117)

One need not claim Voltaire as a feminist[11] to acknowledge the significance of the old woman's position; as the voice of experience, she is wiser than both Pangloss (who represents optimism) and Martin (who declares himself a Manichean, but may be said to represent pessimism generally). Because the tale ultimately recommends experience over theory, as both the vehicle for and the conclusion of Candide's enlightenment, Voltaire's symbolic use of the old woman seems less like misogyny and more like shrewd observation. In the eighteenth century and today, most of the world's women suffer some form of gender-related oppression. In many countries women and female children are still bought and sold, and rape remains a preferred strategy for non-combatant "subjugation" in conflicts across the globe.[12] *Candide* does not condone this oppression as part of 'the way of the world.' On the contrary, Candide is highly sympathetic to the plight of the prostitute Paquette, and to the sufferings of Cunégonde and the old woman (all of whom will become valued members of the Candide's little commune). Yet, ultimately, the crucial gesture for women lies in the text's assertion that the experience of material life has more philosophical validity than metaphysics (which, of course, was an almost entirely male pursuit in the eighteenth century). And, of course, it was not Pangloss or Martin but the old woman "who suggested to Candide that they settle into it [a little farm in the area] until the destiny of the group improved" (*Candide* 115).[13]

The old woman's question (what is worse—horrific physical suffering or philosophical arguments, debates, and boredom?) receives a number of answers in the text. The question leads Martin and Pangloss to philosophize, and after the arrival of the miserable Paquette and Giroflée (further confirmation of Pangloss's philosophical error), Pangloss, Candide, and Martin seek the advice of "a very famous dervish who was reputed to be the best philosopher in Turkey" (*Candide*

---

[11]   In "Voltaire's Candide: A Tale of Women's Equality," Arthur Scherr claims that one of *Candide's* primary themes is the equality and interdependence of the sexes. Gloria M. Russo's "Voltaire and Women," from *French Women and the Age of Enlightenment*, makes a similar and compelling claim for "La Pucelle" and the contes philosophique based on textual and biographical evidence.

[12]   There are a variety of statistics on these subjects. MADRE (an international women's rights organization), International Association for Feminist Economics, United for a Fair Economy, and the UN Report on Women's Progress provide a wealth of information.

[13]   The call for experience in place of theory in the eighteenth century has been considered more reactionary than radical. In Voltaire's narrative, however, experience isn't championed over theory but over empty abstractions. Indeed, *Candide* implies that "the good life" is inseparable from reflection.

117). Not only does the dervish have no answer to the question of human life and its relation to the world, he is utterly indifferent to the question, and hostile to the curious. "'Why meddle in that?' said the dervish, 'Is it any business of yours?'" (*Candide* 118). He also has nothing to say about how to live in a world of suffering, save, "'Keep silent'" (*Candide* 118). On the way back to their farm, the three encounter a kindly old farmer resting in his orange grove; they have just heard rumors of the murders of several Constantinople court officials. Hoping for more information, they ask the old Turk the name of the strangled mufti (law expert), and he replies, "'I have no idea. ... I assume that in general those who meddle in public affairs perish, sometimes miserably, and that they deserve it. But I never think about what people are doing in Constantinople. I am content to sell them the fruits of the garden that I cultivate'" (*Candide* 188).

Having been presented with these inadequate approaches to life (the dervish's attempt to transcend life and the Turk's insular materialism), Candide finds an answer to the old woman's question from his own experience (we will come back to this experience in a moment). Neither physical suffering nor the frustrated pursuit of philosophical ideals is more painful than the other; in fact, *Candide* demonstrates that they exacerbate each other. Candide's estrangement from the world he wanders proceeds not from the faculty of reason but from the inadequacy of metaphysics in the face of calamity. For example, in chapter twenty-three, Candide asks the reason for the Admiral's execution.

> "It's because," came the answer, "he didn't kill enough people. He was engaged in a battle with a French Admiral and was later judged to have kept too great a distance from the enemy."

> "But," said Candide, "the French Admiral was as far from the English admiral as the latter was from the former."

> "That's incontestable," was the response. "But in this country they think it's good to kill an Admiral from time to time, to encourage the others."

> Candide was so stunned and shocked by what he saw and heard that he would not even set foot on land. (*Candide* 99)

At the end of the tale, Candide's full cognizance of the futility of Pangloss's systematic rationalism and other brands of metaphysics (Martin's pessimism and the dervish's mysticism) leads him to create his communal garden. His answer to the old woman is clear: "we must cultivate our garden" (*Candide* 119).

Candide's choice of the thoughtful materiality of the garden over empty abstractions makes sense after all the unhappiness they have caused him and the rest of the world, particularly in the forms of patriotism and religion. Upon returning from the Turk's home, Candide proclaims, "That kindly old man seems to have made a better life than the six kings we had the honor of eating supper with" (*Candide* 119). One might ask why Candide doesn't set up a garden like the Turk's. The answer, I believe, is in the comparison. The Turk's garden is not unlike the courts and noble households we see in *Candide*—insular, selfish, uncaring, and uncompromising; it is, ultimately, antisocial.[14] Yet, Candide clearly sees the benefit of the Turk's particular practical activity—keeping away the "three great evils: boredom, vice, and indigence" (*Candide* 119). If we consider the practice of metaphysics to be a vice, or at least a cause of vice, then Candide's "little society" seems set up as an antidote to transcendental ideas and—despite the humorous tone of the passage—the very serious problems they cause. Everyone participates in the collective freedom from the three evils:

> Each began to exercise his talents. The little bit of earth became productive. Cunégonde was undeniably very ugly, but she baked excellent pastries. Paquette embroidered. The old woman took care of the linen. No one failed to contribute, not even Brother Giroflée. He was a very good carpenter and even became a sociable fellow. (*Candide* 119)

Unlike a king's court or the Turk's garden, Candide's "little society" functions like a commune—a niche in which each contributes according to ability for the welfare of the group. *Candide's* portrait of the good life is one in which individuals function together for their mutual aid.

Jardinier

Candide suggests that a conscious return to the land, as the practice of material awareness, remains the best course for social happiness. Certainly, Voltaire believed this idea enough to practice it in his own life. "Voltaire," writes Raymond Williams, "saw the pursuit of industry and urbane pleasure as the marks of the city and thence of civilization itself. The golden age and the Garden of Eden, lacking industry and pleasure, were not virtuous but ignorant: the city, especially London, was the symbol of progress and enlightenment" (*The County* 144). If, as Williams argues, Voltaire identified civilization with "the city," then his move to the country becomes even more deliberate. Perhaps the city, as the seat of industry and commerce, was regarded as a bearer of ill as well as good. Power concentrated in the cities, and with it, corruption and oppression.

Although for many years Voltaire's life was, like Candide's, one of movement—from one part of Europe to another, as courtier and exile—Voltaire eventually made his way to Switzerland and then back to nearby France, where he purchased several country properties over a number of years. The first was *Les Délices*

---

[14]   Voltaire's chief argument against Liebnizian optimism is that it is fundamentally quietist, and so antisocial. In a letter to Elie Bertrand on February 18, 1756, he writes, "Optimism is despairing. It is a cruel philosophy under a consoling name. ... We will go from misfortune to misfortune to become better off. And if all is well, how do Leibnizians allow for something better?" (Brooks 183).

(House of Delights), just outside Geneva, where he resided during the summer and fall of 1758, when *Candide* was probably composed.[15]

An examination of excerpts from Voltaire's letters from *Les Délices* and Ferney, Voltaire's second and favorite farm (he eventually sold *Les Délices*, but he kept Ferney, where he lived until just shortly before his death in 1778), reveals his material commitment to gardening and rural community life. Although gardens, and the pastoral life in general, were in vogue in the eighteenth century, Voltaire's life as a *jardinier* was not for show. Douglas Chambers's *The Planters of the English Garden Landscape* provides an excellent context for Voltaire's gardening:

> [T]hroughout the first half of the eighteenth century, much garden design and writing began with a reaction against the artificialities of both Dutch and French garden design. In doing so its celebration was of a landscape that aspired toward the condition of an ideal farm as Pliny [and Horace describe] ... and Virgil in lines 458–540 of *Georgics* II. Another much-cited tag from the *Georgics*, *laudato ingentia rura, exiguum colito* (praise a large farm but cultivate a small one), became the hallmark of the *ferme ornée*. That most Augustan invention represented the unity of beauty with profit and use with pleasure that was within the means of a man of modest income: the smallholding of a man of philosophical mind. (6)

Voltaire's gardens could easily be described as *ferme ornée* as they were productive instead of merely ornamental. From *Les Délices* on March 28, 1775 (a few months after taking possession of the property), Voltaire writes to Tronchin,

> My thanks for the lavender. I have promised you to have some planted round the edges of your kitchen garden. I have already planted 250 trees for you ... I am now sowing Egyptian onions. The Israelites did not love them more than I. Whereupon I abuse your kindness. I beg you to send me all that you can in the way of flowers and vegetables [seeds]. The garden is absolutely lacking in them. (*Select* 147)

Voltaire's reference to "your kitchen garden" is a little joke; Tronchin had to register as the legal owner of the property because in Calvinist Geneva, a Catholic (even a lapsed one) could not own property. This letter, and dozens more like it, demonstrates Voltaire's interest in and regular interaction with his garden.

While Voltaire's work in the garden begins at *Les Délices*, his enthusiasm blossoms at Ferney, in nearby France. From the *Chateau de Ferney* on August 1, 1772, Voltaire writes to Sir William Chambers (who had recently sent Voltaire a copy of his monograph, "Dissertation on Oriental Gardening"):

---

[15] There are varying opinions on this. In *Voltaire's* Candide: *the Protean Gardener, 1755–1762*, Geoffrey Murray advances this claim, while Theodore Besterman believes that he may have been staying at Schwetzingen when Candide was composed (24–7; 408).

> It is not enough to love gardens, and to have them: one also needs eyes with
> which to look at them and legs with which to walk in them … I have something
> of everything in my gardens: flower beds, little pieces of water, formal walks,
> very irregular woods, valleys, meadows, vineyards, kitchen gardens with walls
> covered by fruit trees, the groomed and the wild, the whole in little. (*Select* 169)

Groomed and wild, "the whole in little"; this garden isn't a miniature Eden but
a reflection of the interconnectedness of the human and nonhuman, microcosm
and macrocosm. In *The Life of Voltaire*, S.G. Tallentyre records that although the
grounds at Ferney were rather elaborate, containing acres for wheat and hay, poultry
yards, sheepfolds, an orchard, beehives, vegetable gardens, and more, Voltaire not
only oversaw every detail of this garden, he cultivated a good bit of it with his own
hands, which were no longer so young (86). Tallentyre describes Ferney thus: "In
the garden were sunny walls for peaches; vines, lawns, flowers. It was laid out with
a charming imprévu and irregularity most unfashionable in that formal day" (86).
From Voltaire's letters and this description, we see that he had indeed abandoned
the Dutch and French fashion for formal, purely decorative gardens in favor of a
somewhat wilder, productive *ferme ornée*. The intentional informality of Voltaire's
gardens, both in *Candide* and at *Les Délices* and Ferney, seem particularly
important in light of Pope's *Essay on Man*, to which, of course, *Candide* was a
rejoinder every bit as much as Leibniz's treatises. Chambers summarizes Pope's
metaphoric use of landscape in terms which make the comparison clear: "In the
words of Pope's *Essay on Man*, the universe is a landscape garden: a mighty maze,
but not without a plan" (5). Voltaire once said: "I have only done one sensible thing
in my life—to cultivate the ground. He who clears a field renders a better service to
humankind than all the scribblers in Europe" (Tallentyre 87).[16]

Perhaps even more importantly, Ferney was not an estate, such as *Les Délices*,
but a community. In his autobiography, written in the third person, Voltaire
observes,

> [T]he village of Ferney, which at the time of his purchase, was only a wretched
> hamlet tenanted by forty-nine miserable peasants, devoured by poverty, scrofula
> and tax-gatherers, very soon became a delightful place, inhabited by twelve
> hundred people, comfortably situated, and successfully employed for themselves
> and the nation. (qtd. in Besterman 662)

Indeed, one of the chief reasons why Voltaire purchased the land was to alleviate
the tremendous suffering of the people of the region; he confided in a letter to a
French correspondent, "One's heart is torn when one witnesses so much misery. I
am buying the estate of Ferney only to do a little good" (qtd. in Besterman 407).
Numerous biographers, including Besterman, Tallentyre, and Norman L. Torrey,

---

[16]   This seems an allusion to a famous passage in chapter seven, part two of *Gulliver's
Travels*: "whoever could make two ears of corn, or two blades of grass, to grow upon a
spot of ground where only one grew before, would deserve better of mankind, and do more
essential service to his country, than the whole race of politicians put together."

catalogue Voltaire's activism on behalf of this little society (including financial support; the removal of the local *curé*, who ruthlessly beat his parishioners; the drastic reduction of taxes; and the construction of roads and communal buildings).

Geoffrey Murray's *Voltaire's* Candide*: the Protean Gardener, 1755–1762*, repeatedly suggests that Voltaire used his "garden" as a detached vantage point from which to observe the horrors of the Seven Years War (153–221).[17] While Murray claims, in so many words, that Voltaire was an escapist who justified his intellectual pursuits with humanitarian rhetoric, both *Candide* and Voltaire's work at Ferney suggest the opposite: purposeful engagement with the world. For Voltaire as for Candide, the good life must be lived ethically, purposefully or, as Henry David Thoreau put it, deliberately. And like Thoreau's little garden at Walden Pond, from which *Walden* grew, Candide's "little bit of earth" and Voltaire's gardens "became productive" (*Candide* 119).

Technology and the Bioregional Land Ethic

As ecologists and activists have been saying for decades, technological "advancements" and global "development" usually mean the maldevelopment of entire bioregions, the destruction of flora and fauna and ways of life.[18] In his call for self-critical reason, Racevskis comments on the systematic rationalism of "the contemporary technological arrangement":

> This perspective [the Kantian idea that social progress accompanies new knowledge] also shows that the possibility of placing knowledge outside the conditions that produced it … constitutes a mechanism that is fundamental to the contemporary technocratic arrangement. Thus, it is evident that attempts to rationalize society in order to make it transparent to the scientific, technical mind are still driven by a desire to fulfill the telos of *an* Enlightenment-oriented ideology. The technical expert has given rise, for example, to the "heroic expert," the incarnation of the dominant instrumental ethic, which is also a dominant masculinist ethic, as a number of feminists have argued. (83, italics mine)

Indeed, as Maria Mies argues in *Ecofeminism*, since the time of Bacon and Descartes, scientists "have consistently concealed the impure relationship between knowledge and violence or force (in the form of military power, for example) by defining science as the sphere of a pure search for truth. Thus they lifted it

---

[17]   Murray also connects *Candide* with Voltaire's life; his book catalogues Voltaire's progressive interest in agriculture and animals with the purchase of *Les Délices* in 1755. Despite his conclusions, I agree with Renee Waldinger's assessment of the book: "Murray demonstrates convincingly that the language of this [Voltaire's] correspondence is the very language of *Candide* and that the final garden is not an abstraction but a real spot on earth where productive life is possible" (17).

[18]   There are a number of accounts of this process. See Carolyn Merchant's *The Death of Nature* for a historical account and Maria Mies and Vandana Shiva's *Ecofeminism* for a theoretical overview.

out of the sphere of politics ... [T]he separation of politics (power) and science ... is based on a lie" (46).[19] Donna Haraway's response to this lie is a call for a "successor science," which would take "account of radical historical contingency for all knowledge claims ... [while making] a no-nonsense commitment to faithful accounts of the 'real world'" (*Simians* 187).

The bioregional land ethic, or gardening model,[20] stands in opposition to the universalizing, instrumental rationalism of technological capitalism through the commitment to the lived stewardship of particular places. Voltaire often claimed that he was a citizen of all nations and a citizen of no nation, a citizen of the world and a resident of the *pays de Gex* and Geneva (Murray 72). His cultivation of *Les Delices* and Ferney was a rejection of national systems in favor of a kind of regionalism, a commitment to a particular place, just as Candide's garden suggests a community of care. A bioregional ethic approaches the successor science Haraway imagines; it is an embedded, contextual basis for knowledge claims and a commitment to the real world. The garden model seeks to use local knowledge and experience to care for habitats and their inhabitants simultaneously; in this way, communal "gardening" might embody the ethic of care and responsibility Plumwood advocates. Unlike the rationalist models of mainstream science, successful bioregional stewardship (or "gardening") requires a respect for the lived interconnectedness of human and nonhuman natures (and cultures).

"'I know,' said Candide, 'that we must cultivate our garden.' 'You are right,' said Pangloss, 'for when man was placed in the Garden of Eden, he was placed there *ut operaretur eum*, in order to work on it, which proves that humankind was not made for rest'" (*Candide* 119). As *Candide* indicates, the garden is not without its dangers. This model may become yet another form of "management" that privileges the human at the expense of the more-than-human world, like the Judeo-Christian edict Pangloss invokes.[21]

The idea of wilderness has long functioned as a guiding metaphor for ecological thought; yet in recent years, many critics have challenged wilderness as a metaphor or model, and rightly so, on the grounds that it has served to "erase"

[19]   The real political power of this lie problematizes even aspects of ecology. This has been the subject of much debate in ecological circles. Many deep ecologists, social ecologists, and ecofeminists feel that the entire program of "land management," including conservation reserves, is complicit in imperialist land-grabs, corporate "greenwashing" campaigns, the commodification of ecological activism, and the perpetuation of dualistic thought which defines nature as "out there." Other ecological thinkers maintain that land, air, and water management have never been more necessary.

[20]   While some ecological thinkers may claim differences between the bioregional movement, Aldo Leopold's land ethic, and "gardening" as an ecological model, I believe they basically advocate the same thing: the embedded stewardship of particular places as ethical practice and ecopolitical activism. And so I have collapsed bioregionalism and the land ethic into bioregional land ethic, which I use interchangeably with the concept of the "gardening model."

[21]   Frederick Turner's garden model is a particularly telling example of this danger; see "Cultivating the American Garden" in *The Ecocriticism Reader* or in Turner's *Rebirth of Value: Meditations on Beauty, Ecology, Religion, and Education*.

human presences from the landscape in the service of economic exploitation (as in the American West), and continues to reinforce a conception of nature as static and separate from humanity. For these and other reasons, some critics of wilderness argue that gardening should become the central metaphor of ecology. But, again, this idea carries risks too. Gardening as a guiding principle may elide the limits of human knowledge and power. It can slip into a naturalization of hierarchy, in which human gardeners are the pinnacle and purpose of creation. It can absorb everything, leaving no room for the wild. We are all on this planet together, but that does not mean that we can all thrive in the very same space. By recognizing that not all beings are compatible in the same niche—that certain animals will always require places free from human encroachment—the idea of wilderness presupposes limits on human power and action. A "wild" garden, one which embraces the animality of the gardener, the politics of the garden, and the wildness of culture itself, may be a way to conceive of the coexistence of different but connected natures.

An article on "extreme gardening" in *The New York Times Magazine* relates the experience of two professors who created a garden tended by "smart machines" that could be manipulated through the Internet (Nussbaum 24). The professors explain that the idea was intended as a short-term "dystopian vision"; however, "over the last three years, 30,000 people have planted seeds via modem. 'It gets pretty ugly in there,' sighs Goldberg. 'On the Web site, there's a quote from Voltaire, which basically says, tend your own garden. A lot of people seem to miss that.'" Between living deliberately in nature and the mastery of it is, one might say literally, all the difference in the world. Candide's injunction to cultivate our garden only makes sense in the context of a lived skepticism toward systematic rationalism in all its guises, including speciesism. It is not sufficient to philosophize like the dervish, in abstraction from the meaning of material life, and it is not enough to garden like the Turk, with no care for the rest of the world. Embracing philosophy as a lived praxis, Candide's cultivation avoids the transcendentalism of the dervish, the logical extension of Pangloss's systematic rationalism, and the individualism of the Turk, the logical extension of Martin's creed, "Let us work without theorizing" (*Candide* 119). And it avoids a naturalization of hierarchy among humans— wisdom that, one hopes, might extend to all of nature.

Dialectic of Enlightenment

Dialectic of Enlightenment is, in a sense, an ecocritical *tour de force*. Horkheimer and Adorno's critique of rationalism and technology argues that, in contrast to the Enlightenment claim to liberate humanity, "the wholly enlightened earth is radiant with triumphant calamity" (*Dialectic* 1). Their great insight into the logic of domination is that domination is, fundamentally, the domination of nature through a species of logic. Their analysis of this logic uncovers Enlightenment's dialectical relationship to myth, which it sought to exterminate through the hegemonic application of rationalism to the world, thus reverting to mythology itself. They argue, "Enlightenment has always regarded anthropomorphism, the projection of subjective properties onto nature, as the basis of myth. The supernatural, spirits and demons, are taken to be reflections of human beings who allow themselves to be frightened by natural phenomenon" (*Dialectic* 4). Like the great rationalists of seventeenth-century thought, Horkheimer and Adorno also view an animate, agential nature as a human construction. In fact, the idea of nature itself becomes a measure of Enlightenment's failure: like myth, Enlightenment's view of nature as a system of laws enacts the determinism it supposedly seeks to escape (*Dialectic* 8).

Adorno identifies both positivism and ontology ("the mythology of Being") as reified consciousness precisely because they claim to know nature: "The more reified the world becomes, the thicker the veil cast upon nature, the more the thinking weaving that veil [positivism and ontology respectively] in its turn claims ideologically to be nature, primordial experience" (*Critical Models* 7). Looking at Plumwood's *Feminism and the Mastery of Nature* or Mellor's *Feminism and Ecology*, one can see that Adorno's critique of positivism, as Voltaire's critique of rationalism before it, laid the groundwork for the ecological critique of rationalism.[22] However, Adorno's attack on ontology misses its radical edge, the dramatic yearning of its struggle to do the most important thing philosophy can—engage with the world. By positing the reciprocity inherent in perception, ecologically-minded phenomenology attempts to offer hope for the world through an enriched experience of the world. David Abrams,[23] for example, argues that human and nonhuman natures contact each other in the act of perception itself and that, when experienced actively, self-reflexively, this perception is a kind of communication between extremely different sites of experience or, to put it differently, between species of subjectivity.

While Horkheimer and Adorno rightly insist that the "mastery of nature draws the circle in which the critique of pure reason holds thought spellbound" (*Dialectic* 19), they take this assertion to a terrifying end:

> However, this thought, protected within the departments of science from the dreams of a spirit-seer, has to pay a price: world domination over nature turns against the thinking subject itself; nothing is left of it except that ever-unchanging "I think," which must accompany all my conceptions. Both subject and object are nullified. The abstract self, which alone confers the legal right to record and systematize, is confronted by nothing but abstract material, which has no other property than to be the substrate of that right. The equation of mind and world is finally resolved, but only in the sense that both sides cancel out. (*Dialectic* 20)

While Horkheimer and Adorno posit this condition as a false equation, they present this profound alienation as historically inevitable and practically inescapable. There are, however, moments when *Dialectic* may be read as less sealed off:

---

[22]   Plumwood considers Horkheimer and Adorno as foundational critics of rationality and the purely instrumental use of nature (24–5). However, in *Wild Knowledge*, Will Wright puts forward an extended criticism of Horkheimer and Adorno's inability to escape the rationality of scientific assumptions and embrace a "social-natural" critique of science (148–53).

[23]   See *The Spell of the Sensuous*, especially the chapter "Philosophy on the Way to Ecology."

> The technical process, to which the subject has been reified after the eradication of that process from consciousness, is free of the ambiguity of mythical thought as from meaning altogether, since reason itself has become merely an aid to the all-encompassing economic apparatus. Reason serves as a universal tool for the fabrication of all other tools, rigidly purpose-directed and as calamitous as the precisely calculated operations of material production, the results of which for human beings is beyond all calculation. (*Dialectic* 23)

Here we catch a glimpse of something moving behind the frame, "the results" of human activity. While the subject, meaning the human subject, objectifies itself in the processes of technology, nonhuman nature, even at its most elemental, cannot be rendered a passive object, a static background. Nature acts against its instrumental use with ecological catastrophe. The results of climate change may well be *beyond all calculation*, but they are not necessarily beyond imagination.

Experience of the world, on which both imagination and thought depend, constitutes our hope. Horkheimer and Adorno emphasize the inextricable connection between experience and thought in the process of domination:

> This regression [of progress] is not confined to the experience of the sensuous world, an experience tied to physical proximity, but also affects the autocratic intellect, which detaches itself from sensuous experience in order to subjugate it. The standardization of intellectual function through which the mastery of the senses is accomplished, the acquiescence of thought to the production of unanimity, implies an impoverishment of thought no less than of experience; the separation of the two realms leaves both damaged. (*Dialectic* 28)

While this analysis identifies the domination of the senses as the cause of both the impoverishment of thought and experience, it does not discuss these realms as shared processes, shared with other creatures, other social worlds, other sites of experience. If the "whole world is passed through the filter of the culture industry," causing real life to become indistinguishable from capitalism's products (99),[24] then perhaps the best hope of change within the human social world is from "without," from contact with other, overlapping social worlds. An inability to imagine truly *other* social worlds—an agential nature of nonhuman cultures— seems symptomatic of this manufactured alienation. The loss of the imaginary and the real goes hand in hand.

Because Horkheimer and Adorno do not seem to allow for the possibility of our experience of the world as a shared process, the human species seems isolated,

---

[24] Horkheimer and Adorno's analysis of cartoons provides an example (138). Seemingly unreal, cartoons come to constitute reality, especially for their primary audience. The authors link this to violence, but do not discuss the significance of the fact that almost all cartoon characters are nonhuman animals. These animals, humans in nonhuman form, form a bridge between signifiers of the human and nonhuman world, allowing culturally sanctioned (even required) cruelty to other animals to metamorphose into culturally sanctioned cruelty to other humans.

its existence solipsistic. In fact, they go so far as to assert that, in "a certain sense, all perception is projection" (154). While positivism and certain ontologies undoubtedly reify (and are reified) consciousness, *Dialectic*'s epistemology treats the human subject as the *only* subject—a virtual prisoner, without the consciousness possible to initiate change or the possibility of experience that changes. With the earth, for all intents and purposes, emptied of all other subjectivities, all cultures save our own, the world seems emptied of hope itself.

In "Mammoth," number seventy-four of *Minima Moralia: Reflections from Damaged Life*, Adorno elucidates a key aspect of his dialectical critique of Western culture: "The more purely nature is preserved and transplanted by civilization the more implacably it is dominated ... The rationalization of culture, in opening its doors to nature, thereby completely absorbs it, and eliminates with difference the principle of culture, the possibility of reconciliation" (115–16). In *Dialectic*, culture seems like fast-drying concrete, invulnerable (or nearly invulnerable) to the transformative power of weeds. Yet in *Minima*, published after *Dialectic*, the "principle of culture" is itself the possibility of reconciliation with nature—though the "rationalization of culture" has the opposite (and dominant) effect. Arguably, it is this principle of culture, which stands in opposition to its rationalization, which the findings of cultural biology revive.

Anthropocentric solipsism, the resignation of the human species to solitary confinement on Earth, is posited as a basic condition of existence, as "natural fact" instead of cultural construction (and a carefully guarded one at that). As Horkheimer and Adorno acknowledge, "Throughout European history the idea of the human being has been expressed in contradistinction to the animal. The latter's lack of reason is the proof of human dignity" (203). However, they affirm the idea that these other creatures have no real thought, no subjectivities: "The animal responds to its name and has no self ... one compulsion is followed by another, no idea extends beyond it. ... In the animal's soul the individual feelings and needs of human beings are vestigially present, without the stability which only organizing reason confers" (205).

In his meditation on the relationship between nature and culture in *Metaphysics*, Adorno defines culture (that is, human cultures) as the attempt to control nature: "everything we call culture consists in the suppression of nature and any uncontrolled traces of nature" (118). He identifies culture's failure as the inability to control death, "the failure of culture stems from its own naturalness ... of its own persistent character as a natural entity" (129). Here Adorno gives foundational weight to the material world, but in wholly negative terms—the inevitability of death. Adorno wants to rescue culture as a transcendent entity, even though that rescue too can only be negative (although culture has failed what takes its place, "barbarism," is worse). While Adorno repeatedly demonstrates that human cultures mediate nature, he doesn't acknowledge that nonhuman natures mediate—even create—human cultures. Culture is a "natural entity," but in *Dialectic* that naturalness seems little more than a Freudian nightmare of shit and death, a repression that eternally returns as the threat of barbarism. The field of cultural biology suggests the reverse—that the naturalness of human culture, of culture itself, may be our best hope. It reminds us that we are part of a larger agential community, reframing mediation as a multidirectional complex of co-creating nonhuman and human cultures.

Unexpectedly, an idea of community with nonhuman nature emerges from one of the earliest examples of the genre of individual development. *Candide* removes the subject from his "origin" in metaphysical speculation and places him in cooperative community, coalesced by a shared desire for the good life: a life of right relations with the world, a life eschewing violence and oppression. "'Power and glory,' said Pangloss, 'are very dangerous, as all the philosophers tells us. ... 'I know,' said Candide, 'that we must cultivate our gardens'" (119). Here, cultivating our gardens means revisiting our assumptions about nature and culture, about the subjectivity and agency of other beings. Back in chapter sixteen, Candide and Cacambo witness two nude women pursued by monkeys, "snapping at their buttocks" (73). Concerned for the women, Candide raises his musket and shoots the monkeys, whereupon the women "tenderly embrace the two monkeys, burst into tears over their bodies, and fill the air with cries of intense grief" (73). Cacambo then informs the bewildered Candide that he has just shot their lovers, to which he replies, "You must be joking. What evidence do you have?" (73). But this is no joke, as the intense cries of the women should have already made clear. Annoyed at his incredulity, Cacambo asks, "Why do you find it so bizarre that in some countries monkeys obtain the favors of the ladies? They are a quarter human, just as I am a quarter Spanish"—and Candide answers, "I do recall hearing Dr. Pangloss say that such things used to happen, that these mixtures engendered pans, fauns, and satyrs, and that many heroes of ancient times saw them. But I thought that was all a fable" (73).

Pans, fauns, and satyrs, yes, but it isn't *all* a fable. The stuff of myth is, in chapter sixteen, more than an animate nature; it is the reality of nonhuman subjects living in active community with human subjects. Though perhaps the form of this *active* community is, like much else in *Candide*, more than a little exaggerated (interconnection becomes intercourse), this chapter suggests that some of what rationalism treats as myth—in this case, nonhuman subjectivity, the multi-species nature of community—is real. Cacambo's statement that monkeys are "a quarter human" just as he is "a quarter Spanish" is also interesting; it supposes, before Darwinian theory and long before genetics, that animal life is truly continuous.[25] In light of Cacambo's next statement, it may also be read as a witty critique of the "educated" animalization of non-Western peoples (here, Cacambo's own "three-quarters" indigeneity[26] and the "savages called Oreillons"), rather than an

---

[25] In *Literature After Darwin*, Virginia Richter writes, "The search for the missing link ... becomes the El Dorado of post-Darwinian debates, both in the hard realities of palaeoanthropology and in the imaginary space of literature" (44). It is interested that Voltaire follows Candide's discovery of the monkey-lovers with his discovery of El Dorado.

[26] Candide takes Cacambo on as a valet in Cadiz. The editor, Daniel Gordon, footnotes the following: Cacambo's "mother was a South American Indian, his father was half Indian, half European." (68).

affirmation of the racist Chain of Being or Panglossian "metaphysico-theologico-cosmolo-boobology" (42). He replies to Candide: "Now you should be convinced that it's the truth ... and you can see how this truth is still practiced by people who have not been educated" (73).

The Oreillons capture Candide and Cacambo for the murder of the monkeys and begin preparations to eat them. Candide asks Cacambo, as he is about to try to communicate with the Oreillons, "'Remember to try to explain to them,' added Candide, 'that it is the height of inhumanity to cook human beings, and that it's not very Christian either'" (74). The two escape being eaten with the acknowledgment that, while it makes good sense to eat one's enemies rather than "abandon the fruits of victory to crows and ravens," they are not in fact their enemies; despite Candide's appearance as a Jesuit, he is not one (Candide is wearing the robe of the Jesuit he killed in the previous chapter). Candide concludes this interlude with a characteristically "candid" observation, "it turns out the pure state of nature is good, because I only had to show these people I wasn't a Jesuit, and they treated me with enormous kindness" (75).

Along with the lampooning of Jesuits, chapter sixteen presents, at once, a critique of rationalism, as the naïve assumption of human superiority or separateness, and a poke at the naïve idealization of nature, in the form of Candide's silly remark about the goodness of "the pure state of nature." The link between mistaken ideas of nature (human and nonhuman), exploration, animalization, and exploitation is dramatized in the sequence and content of chapters sixteen[27] through nineteen. Sixteen is followed by Candide and Cacambo's discovery of the mythical El Dorado in chapter seventeen; chapter eighteen also takes place in El Dorado, where they stay for one month before arriving in Surinam in chapter nineteen (where Candide converses with a sugar plantation slave, in a passage quoted earlier in this chapter). El Dorado, the "golden one," is a utopia not because of its abundance of gems and gold—but because of its social and material harmony. Surinam, despite, or perhaps because of, its "natural resources" (that is, the designation of human and nonhuman nature as a resource) is quite the opposite.

While *Candide* and its garden do not offer a developed critique of speciesism, the narrative does posit the radical, materialist skepticism necessary for one. Many of the ideas present in *Candide* will return in Chapters 5 and 6, which explore among other things the relationship between animalization and dehumanization (and Empire and globalization), prefigured in the monstrosity of an agential nonhuman nature discussed in the next chapter.

---

[27] Daniel Gordon notes: "'Oreillons' means big ears. In Garcilasco de la Vega's *Comentarios reales*, a work first published in 1609, Voltaire had read about a tribe of Indians in Peru who pierced and distended their ears. According to Vega, the members of this tribe had a propensity for nudity, sodomy with monkeys, and cannibalism" (74). See Richard Brooks's "Voltaire and Garcilasco de la Vega" in *Studies on Voltaire and the Eighteenth Century*.

Chapter 3
Ecocriticism and the Production of Monstrosity in *Frankenstein*

I do not know that the relation of my misfortunes will be useful to you, yet, if you are inclined, listen to my tale. I believe that the strange incidents connected with it will afford a view of nature, which may enlarge your faculties and understanding.

—Victor Frankenstein, *Frankenstein*

Thou hast a voice, great Mountain, to repeal
Large codes of fraud and woe; not understood
By all, but which the wise, and great, and good
Interpret, or make felt, or deeply feel.

—"Mont Blanc," Percy Bysshe Shelley

Early in *Civilization and its Discontents* Freud writes, "Voltaire has deflections in mind when he ends *Candide* with the advice to cultivate one's garden; and scientific activity is a deflection of this kind, too" (24). Freud reads *Candide*'s conclusion as one of negative resignation, another "palliative measure" for life's pain and disappointment, a deflection "to make light of our misery" (23). This misreading of Candide's assertion follows another key misreading in humanism's construction of systematic rationalism, that of the "oceanic feeling," a phrase Freud derives from the term by which his correspondent, Romain Rolland, described the sensation of the world as limitlessness or eternal, "unbounded—as it were, 'oceanic'" (11).

This sense isn't an impression of personal immortality; rather, "it is a feeling of an indissoluble bond, of being one with the external world as a whole" (12). Freud analyzes this as a sensation of the externalization of the ego (or primary narcissism), normally associated with a variety of pathological conditions or the ecstasy of love, an echo of an earlier state of the psyche before the defense mechanism of differentiation:

Our present ego-feeling is, therefore, only a shrunken residue of a much more inclusive—indeed, an all-embracing—feeling which corresponded to a more intimate bond between the ego and the world about it. If we may assume that there are many people in whose mental life this primary ego-feeling has persisted to a greater or lesser degree, it would exist in them side by side with the narrower and more sharply demarcated ego-feeling of maturity, like a kind of counterpart to it. In that case, the ideational contents appropriate to it would be precisely those of limitlessness and of a bond with the universe—the same ideas with which my friend elucidated the "oceanic" feeling. (15)

As evidence for his idea of the residual persistence of the original psyche, Freud turns to evolutionary theory: "In the animal kingdom we hold to the view that the most highly developed species have proceeded from the lowest; and yet we find all the simple forms still in existence to-day" (16). In the mind too, Freud argues, the "primitive" may be preserved alongside the evolved. While he gives us many caveats, accounting for the fact that species alive today are not the direct ancestors of "higher" species (and that intermediate species have died out), Freud's analogy may be more useful than he thinks.

This discussion of the oceanic feeling serves as a bridge between Freud's earlier text on religion, *Future of an Illusion,* and the main argument of *Civilization.* He comes to the conclusion that the oceanic feeling does not present a strong enough claim to be the source of religious feeling because, "a feeling can only be a source of energy [for another feeling] if it is itself the expression of a strong need" (20). Instead, Freud argues for infantile helplessness and the corresponding need for the father as a clearer psychic source of religious feeling, though he digresses into uncertainty and obscurity (as, he claims, does the subject itself). Freud assumes that the oceanic feeling is not the expression of a primary need, perhaps because, as he admits, he can find no trace of this feeling in himself (11).

However, this phenomenon may indeed be read as an expression of a *primary* need, our need for reality, to feel part of the fabric of the world.[1] In this sense, the oceanic feeling serves as an example of what E.O. Wilson calls "biophilia": the innate feeling of our real connection to all forms of life.[2] From this perspective, Freud's recourse to evolutionary theory is telling. In order to argue for the persistence of earlier stages of life in the mind, he looks toward earlier stages of planetary life, suggesting that microcosm and macrocosm are materially interconnected. In fact, the truth in this analogy is, in smaller terms, the primary thesis of *Civilization*: the human mind and human civilization are, for better or worse, co-creating and co-sustaining. The individual psychic system extends beyond itself, belonging to a larger continuum of phenomenon. The oceanic feeling seems an expression of just such continuity. In Freud's own terms, human society, or a sense of personal interconnectedness, is an expression of a primary need (initially expressed through the social unit of the family); it is, then, not such a large step to read the psychic need for interconnectedness in biological terms, as an expression of a need for other forms of life shared by all forms of life.

The oceanic feeling functions as social feeling that extends outwards to both human and nonhuman others. The defended, differentiated self lives alongside this undifferentiated self; if we discard the prejudice that often accompanies the categories of "primitive" and "evolved," there is no reason not to assume that it is this coexistence that expresses the most basic instinct of the human psyche,

---

[1]   In *Writing for an Endangered World,* Lawrence Buell writes that long before contemporary science revealed our origins in the oceans, bodies of water have been a common symbol for primordial reality; he characterizes Freud's use of the word "ocean" as symbolic of "unbounded inner space" as a feeling of eternity (199).

[2]   See *Biophilia* (Harvard UP, 1984).

the interconnected instinct for life as self-preservation, species preservation, and the preservation of other species. Again, this is what Marc Bekoff means when he characterizes anthropomorphism as "an evolved perceptual strategy" (10), arguing that "If we don't anthropomorphize, we lose important information. ... it is a necessity, but it also must be done carefully, consciously, empathetically, and biocentrically. We must make every attempt to maintain the animal's point of view" (124–5).

This oceanic or ecosocial feeling is, to borrow a use of the term from Mary Mellor, an "immanent" sensibility, a sense of the interconnectedness of the world. This sensibility is reified in capitalist culture, transformed into its opposite, into a drive for transcendence and the corresponding denigration of the material world. Immanence (agency and relationality) becomes object ("dead" matter) and teleology. Contrary to Freud, we might indeed see religious feeling as a displacement or deferral or, more accurately, reification of the oceanic feeling, in as much as religion corresponds with transcendence of *this* world. As *Candide* indicates, systematic rationalism (and its current guise, technological rationalism) is another such form of transcendence. The belief that technological development, as a unified generative force, will enable human life to "evolve" indefinitely promotes a view of humans as, paradoxically, inherently outside and above the rest of nature. This view of technology and the material world fuels and is fueled by the enterprises of late capitalism, from monocropping and factory farming to the genetic modification and creation of organisms (such as Craig Venter's new bacteria). As one recent article on Mary Shelley's[3] novel suggests,[4] it is just this kind of technological intervention at issue in *Frankenstein.* One might even say that Victor Frankenstein is an *uncritical* posthumanist.

Both *Candide* and *Frankenstein* problematize systematic rationalism; as narratives of individual human development, they speak to the humanist notion of development itself, and its (technological) trajectory. Chapter 2 examined the way in which *Candide* posits a version of the gardening model, one of interconnectedness and care over one of isolation (or objectification and hierarchy), as the path toward "the good life." This chapter will analyze *Frankenstein*'s dramatization of the costs and consequences of the drive for transcendence in terms of humanist culture's anxieties (conscious and unconscious) about human and nonhuman identities in capitalist production. First I will consider the novel's considerable critical landscape and then its composition history, using a reading of Percy Bysshe Shelley's "Mont Blanc" to open a key moment in Shelley's novel ("Mont Blanc" was also composed, in part, during the summer of 1816 and, like certain passages in *Frankenstein*, first published in *History of a Six Weeks' Tour* in 1817).[5]

---

[3]   Hereafter referred to as Shelley.

[4]   See Teresa Heffernan's "Bovine Anxieties, Virgin Births, and the Secret of Life" in *Cultural Critique* 53, (Winter 2003): 116–33.

[5]   Charles E. Robinson's publication of two manuscript versions of the novel in *The Original Frankenstein* in 2008 demonstrates Percy's editorial influence on what is certainly Mary's text. This close writing relationship is also evident in *History*.

Ecocriticism and Romantic Studies

Jonathan Bate's 1991 *Romantic Ecology* legitimized exploring literature's relationship to ecology. Five years later, P.M.S. Dawson complains that Bate's choice of William Wordsworth as his subject serves as an easy confirmation of ecological values ("'The Empire'" 232–3). Calling for ecocriticism to examine texts that "challenge our beliefs" instead of those that reinforce them, Dawson turns to the work of Percy Shelley as an example that complicates critical impulses and expectations. A large (and still quickly growing) number of ecocritical analyses have taken up this challenge both within and beyond Romantic studies, with the work of Percy Shelley and countless other "complicated" choices in British, American, and Anglo-postcolonial literatures, and beyond.

On the other hand, until the last few years, Shelley had been almost ignored by ecocritics. Bate's brief analysis of *Frankenstein* in *The Song of the Earth* was one of only a handful of ecocritical works on the text. The recuperation of Shelley's work, *Frankenstein* in particular, began in the 1970s with Sandra M. Gilbert and Susan Gubar's *The Madwoman in the Attic* and George Levine's critical anthology, *The Endurance of Frankenstein*. Why, one must ask, has ecocriticism been slow to join the conversation?

Ecocriticism first gained legitimacy in Romantic studies because many of its critical concerns were implicit in the field from its inception. Questions about the relationship between human beings and the rest of nature, knowledge and power, and the various effects of the Industrial Revolution are central to Romantic writing and scholars of Romanticism. *Frankenstein*, perhaps more than any other text of Romanticism, explicitly asks such questions, challenging readers to engage layers of ambivalent theorizing of dialectical relations (between characters, ideas, other texts, and so on). For this reason, *Frankenstein* and the context of its production seem a good case study for an examination of what ecocriticism stands to gain from a more materialist, dialectical approach to literature and culture.

As the critical discourse on and around *Frankenstein* continues to reach staggering proportions (much of it written in the last fifteen years), the novel's ambivalence seems even more pronounced; the critical body magnifies fissures in textual meaning, leaving a labyrinth of relations. Criticism tends to focus on the following thematic categories: birth and the body (feminist/gender and psychoanalytic criticisms), monstrosity and/or the sublime (historicist, linguistic, and psychoanalytic criticisms), politics of creation or the creature/creator dichotomy (linguistic, historicist, identity, and postcolonial criticisms), intertextuality or education (historicist and reader-response criticisms), and science and production (feminist, historicist, and Marxist criticisms). These overlapping categories all touch upon ecocritical concerns, such as the relationships between characters and their environments, the limits of human power, and the costs and consequences of production, yet hardly any of the many articles and books represented by this list examine these subjects in ecological terms.

Like many historicist and intertextual approaches, Bate's analysis situates the novel in the context of Rousseau's *Discourse on the Origin of Inequality* (and, to a lesser degree, *Emile* and *La Nouvelle Héloise*), characterized as "modernity's founding myth of the 'state of nature' and our severance from that state" (42). For Bate, Rousseau's *Discourse* is a political thought-experiment that also performs the cultural work of an imaginative escape from the problems of civilization, much like the myth of the Golden Age or the Garden of Eden (42–9). Moving from the outer to the inner rings of narration, Bate presents *Frankenstein* as the story of two men on "the Enlightenment quest to master nature," in pursuit of two parallel invisible forces, magnetism and electricity (49). These Enlightenment pursuits sever both Robert Walton and Victor Frankenstein from the rest of the world, including family and friends. Interpreting Victor's education as a "fall" into scientific knowledge, a reenactment of European scientific progress from alchemy to chemistry, Bate reads the novel as a calculated rejection of modern science as the mastery of (and separation from) nature: "It is Enlightenment man who invents the natural man; like Rousseau's second *Discourse*, *Frankenstein* uses the state of nature as a heuristic device to critique both Ancien Régime tyranny and the enlightened aspirations of the present" (52). Here, as in *Discourse*, the state of nature is not only a heuristic device but a *fiction*, a product of the Enlightenment in the same way that the Creature (the bodily representation of the state of nature) is a product of Victor's scientific knowledge. The Creature's fall from grace, his destructive power, is in fact Victor's power turned back upon himself: "It is he [Victor] who refuses to give his Creature any name but Monster, who by treating him as a monster turns him into one. The Creature, we may say, is the repressed nature which returns and threatens to destroy the society that has repressed it" (54).

In "Romanticism and Enlightenment," Marshall Brown stipulates that although the conventional account of Romantic writing as a refutation of Enlightenment ideas contains a good deal of truth, a closer examination reveals that much of Romantic writing continues Enlightenment paradigms (27–8). Brown argues that multiple "Romanticisms" and "Enlightenments" form a continuum of values that reveals Romanticism to be the "self-knowledge" of the Enlightenment, even its "fulfilling summation" (38). Similarly, Bate demonstrates the way in which the novel both continues and critiques the Enlightenment project. He concludes that the novel's closing scenes in the Arctic constitute an image of nature's power to resist cognitive division, and so human domination, but also that this resistance is articulated in strikingly Enlightenment terms: "healthy disorientation" in the Arctic (like political and cultural revolution) counters "mastery" (the reign of dogma and the Ancien Régime it supported) (54–5).

Despite the many references to Frankfurt Theory in *The Song of the Earth*, Bate presents an insufficiently dialectical reading of Shelley's novel. For example, Bate's use of the term "Creature" (and only Creature) to refer to the third narrator, instead of "monster" or any of the other names for him that appear in the text,[6] erases the novel's complex dialogue with ideas of creation and production. To refer to the monster as Creature alone flattens the complexity of both the character and the philosophical and material problems he embodies. As Warren Montag suggests, "the monster is a product rather than a creation, assembled and joined together not so much by a man ... as by science, technology, and industry ... whose overarching logic subsumes and subjects even the greatest of geniuses" (388). The Romantic ideology of creation obscures the historical and material contexts of production. In an attempt to respect the monster's agency (or perhaps to give him more agency) Bate calls him Creature; oddly, however, it has the reverse effect, ignoring the radical potential of the monster and text. As Elsie B. Michie demonstrates, the novel consistently and symbolically replaces narratives of creation with narratives of production, even intertextually, as in the replacement of *Paradise Lost* with Victor's laboratory journal (97).

A dialectical approach to the text must consider the movements between creation and production enacted by the multiple "names" humanist culture uses to classify the monster, and its relation to the explicitly dialectical relationship in the text, that of master and slave. Effacing the dictates of the former (the master's classifications) erases the latter, removing the revolutionary potential of the slave and of monstrosity generally. *Frankenstein* is foremost a tale of this culture's fear: revolution is in the air, monstrosity in the weather.

Both *Frankenstein* and *History of a Six Weeks' Tour* were composed, in greater part, during the coldest, wettest summer in Europe in a hundred years. As John Clubbe writes, the unusual weather of 1816, during which Lake Leman flooded, marks *Frankenstein* and *History*.[7] Clubbe ties a moment in the novel, in which

---

6   Refers to the 1818 edition of *Frankenstein* unless otherwise specified. While there has long been a variety of opinions about the respective value of the two primary texts of *Frankenstein*—the 1818 edition and the 1831 revision—Anne K. Mellor makes a very compelling argument for the 1818 text as more coherent and biographically, politically, and scientifically situated (31). Mellor argues that the 1831 revision replaces an organic view of nature with a mechanistic one, repositioning Victor's scientific hubris as the influence and practice of dark arts and rejecting the core values of the earlier text (36). Some critics have characterized these changes as a result of the pressure of the novel's early reception, while others view it as evidence of a growing conservatism (see Tim Morton's comments on the novel's early reception in *A Routledge's Literary Sourcebook on Mary Shelley's* Frankenstein, and Berthold Schoene-Harwood's critical survey of reception in *Columbia Critical Guide: Mary Shelley,* Frankenstein. Also, see Marilyn Butler's essay, "*Frankenstein* and Radical Science"). Regardless of Shelley's reasons for the changes found in the 1831 revision, I will use the 1818 edition as the primary text of the novel, with occasional recourse to the 1831 revision (primarily Shelley's preface to the later edition).

7   See also "*Frankenstein* and Mary Shelley's 'Wet Ungenial Summer'" in *Atlantis*, Donald Olson et al.'s "The Moon & the Origin of *Frankenstein*" in *Skye and Telescope*, and (with respect to Byron's "Darkness") Jonathan Bate's "Living with the Weather" in *Studies in Romanticism*.

Victor traverses the Lake to see a storm, to a real moment in Mary's experience: Victor sees the storm "from precisely the spot in which on 10 June Mary Shelley had originally seen it surging across the waters and had described it in much the same words" (34). Clubbe attributes the role of lightning in the novel, which appears as a creative and destructive force at key moments in the text, to the incredible storms the Shelleys witnessed that summer.

I will take this a step further. The weather also marks the text negatively (unconsciously), through the voice of humanist culture as its horror of nature. From the incredible size and strength of the monster to the significance of his articulations and Victor's irrational response to his gaze, *Frankenstein* registers the human horror—the terror—of nonhuman nature's agency. Terror indeed; the weather during the summer of 1816 was so unusual that across Europe people feared that the end of the world was at hand.[8]

History and "Mont Blanc"

History of a Six Weeks' Tour is the record of two separate trips through France, in July through August of 1814 and in May through June of 1816. Shelley's journal of these trips comprises slightly more than half of the book as part one, and letters written by both Shelleys comprise part two.[9] Percy's "Mont Blanc" (prefaced with "LINES written in the vale of Chamouni") concludes the text. A reading of the poem's last stanza, in particular the question, "And what were thou, and earth, and stars, and sea/ If to the human mind imaginings/ Silence and solitude were vacancy?" (142–4), provides a critical context for the monster's articulation "What was I?"—the dramatic core of *Frankenstein* explored later in this chapter.

Percy Shelley's most seemingly "transcendent" work, "Mont Blanc" has often been read in terms of disembodied experience. While Shelley is, as Harold Bloom argues, engaged in the project of rewriting the Wordsworthian "sense

---

    [8]   Looking back from 1979, *Scientific American* published an article on this mysterious summer titled, "The Year Without a Summer."

    [9]   As a travel narrative, *History* was only one of many such texts on the Alps, which, by 1814, was a very popular destination. In *The Country and the City* Raymond Williams writes, "It is significant and understandable that in the course of a century of reclamation, drainage and clearing there should have developed, as a by-product, a feeling for unaltered nature, for wild land: the feeling that was known at the time as 'picturesque.' It is well known how dramatically the view of the Alps altered, from Evelyn's 'strange, horrid and fearful crags and tracts', in the mid 1640s ... to the characteristic awed praise of mid and later eighteenth-century and nineteenth- and twentieth-century travelers" (128). In "Travel Writing," Jean Moskal examines *History* in the context of the grand tour's newer, more inclusive middle-class form; as an intergenre, travel literature was a mix of fact and fiction designed to sell (243). Yet, as "Revolutionary Tourism," culminating structurally in Percy's "Mont Blanc," Moskal argues that the genre must have had real currency in contemporary politics; here nature is figured "as the repository of a sublimity" that will someday resurface in politics (245–6).

sublime" (Wordsworth's "ALL" that impels and rolls through all things),[10] this does not necessarily mean that Shelley rejects embodied experience for aesthetic fulfillment (this seems more like a description of New Criticism than Shelley's work). Paul Endo's philosophically inflected reading of the poem contends that it is not a symbol of the river but the river itself that appears to the poet, which resists the appropriation of the aesthetic act, the simulacra-world of the cave of the witch Poesy. Like Bloom, Endo views the poem as an attempt to rewrite the Wordsworthian sublime, but he reads this movement as the creation of a kind of anti-sublime. While the first two lines of "Mont Blanc" appear to promise an adherence to the trajectory of M.H. Abrams's formulation of the "Greater Romantic Lyric," the rest of the poem breaks with genre—with the formula of the easy reconciliation with nature through poetic meditation on sensory experience.

As an attempt to comprehend immediacy, "Mont Blanc" resists appropriating nature by refusing to tell us what the mountain means. In the final analysis, though, Endo reads the poem as a noble failure, arguing that in its last section "Mont Blanc" falls out of the anti-sublime by symbolically relating the mountain to a greater whole, to "the secret strength of things."

I would like to propose another option for a materialist reading of the poem. Drawing on Adorno's "On Lyric Poetry and Society" in *Notes to Literature*, I argue that the last section of the poem does not *only* ask an ontological question, but *also* asks, negatively, an anthropological question emanating from the despair of Western culture, a question that challenges the dualistic epistemologies of systematic rationalism. Adorno claims that the lyric poem is the most social when it seems the least so; in other words, the lyric poem's seeming universality—its intense individuality—is in fact its social content imprinted negatively on the artwork. "My thesis," he writes, "is that the lyric work is always the subjective expression of a social antagonism. ... the lyric subject embod[ies] the whole all the more cogently, the more it expresses itself" (45).

"[W]orks of art," as Adorno argues, "give voice to what ideology hides" (39). As the human social world is always embedded in, indeed co-created by, the larger natural world, this view of art allows us to consider responses to nonhuman nature as well as human culture negatively inscribed in cultural works.[11] Again, this critical position suggests that nonhuman nature not only encompasses and impacts human cultures, in ways that we can and cannot see, but that it also might serve as an intervention in culture, in ways that we both can and cannot understand.

Beginning in the third section of the poem, Shelley describes the subjectivity—the "voice"—of nonhuman nature heard, by some, in humanist culture:

---

[10] Bloom writes that "Mont Blanc" transcends the Wordsworthian sense sublime to create a "civilized" sublime, leading to the celebration of the aesthetic work as the negotiation of the duality of mind and matter (see *The Visionary Company*).

[11] However I do not, of course, agree with Adorno's claim in this essay that it is "only through [false] humanization that nature is to be restored the rights that human domination took from it" ("On Lyric" 41).

The wilderness has a mysterious tongue
Which teaches awful doubt, or faith so mild,
So solemn, so serene, that man may be
But for such faith with nature reconciled;
Thou hast a voice, great Mountain, to repeal
Large codes of fraud and woe ... (76–81)

This wider sense of subjectivity informs the poem's fourth section, which meditates upon the interconnectedness of human and nonhuman histories. Shelley's meditation sets up the rhetorical question that concludes the poem in the fifth section:

... The secret strength of things
Which governs thought, and to the infinite dome
Of heaven is as a law, inhabits thee!
And what were thou, and earth, and stars, and sea,
If to the human mind's imaginings
Silence and solitude were vacancy? (139–44)

While Endo and Bloom read this last section as Shelley's positive assertion of the identity of mind and matter, Karl Kroeber, in an equally problematic way, reads it as an assertion of the necessity of mind to make matter meaningful.[12] And, while Kroeber and Bloom read the conclusion of the poem as continuous with the impulses of the first four sections, for Endo the final section signals the difficulty of maintaining an anti-sublime, the difficulty inherent in preventing reason from supplanting sense and imagination.

Following the assertion of the subjectivity of nonhuman nature in the third section and the recognition of the interconnectedness of human and nonhuman histories in the fourth, I suggest the rhetorical question of the last three lines does not only ask *if the world would exist* or *what the world would be* if humanity equated silence and solitude (an absence of human beings) with vacancy (a lack of agents or agency) but also asks, negatively, *what am I* in a society that *does* equate human language with agency and a lack of language with vacancy, with the absence of meaning? The language of the poem resists systematic rationalism by refusing to settle the question of meaning (and the communication of meaning) in human terms—the "still cave of the witch Poesy" and the "mysterious tongue" of the wilderness. The poem asserts that the world is wider than our view of it and more complex than our systems of rationalizing it; the best we can do to understand the world and our experience in it is to practice a form of interpretation synchronous with feeling. As Shelley writes, "Thou hast a voice, great Mountain, to repeal/ Large codes of fraud and woe; not understood/ By all, but which the wise, and great, and good/ Interpret, or make felt, or deeply feel" (80–83).

---

[12] In *Ecological Literary Criticism* Kroeber argues that these last few lines imply that without human imagination "silence and solitude would be vacuous. That is, they would only be what they are physically: negation, absence, isolation ..." (128).

The persona's rhetorical question in the last three lines exhibits the anxiety of the subject in a reified society; the question betrays the social truth that a society that objectifies "natural" or nonhuman subjects will also objectify human subjects. When "Mont Blanc" seems the least cognizant of its social origin, when it apostrophizes the mountain, it is in fact also negatively reflecting the social conditions of its origin—the (false) solipsism of humanist culture. In this case, it is not, as Adorno claims, transcendent and heteronomous to speak the language (or, perhaps, *a* language) of the "nature boy" ("Cultural" 31).[13]

Opening ourselves to a more complex experience of the world, of experience itself, is exactly what Mary Mellor calls "a grounding awareness" of the world. "Mont Blanc" struggles with our awareness of "The everlasting universe of things" as it "Flows through the mind, and rolls its rapid waves" (1–2); it is a struggle with the problem of immediacy itself in a contradictory, reified world. Like the poem, *Frankenstein* is also unwilling to posit the false identity of the human and the nonhuman, or to erase the agency (and suffering) of the nonhuman for humanist culture.

Frankenstein's Nature

> What may not be expected in a country of eternal light? I may there discover the wondrous power that attracts the needle; and may regulate a thousand celestial observations, that require only this voyage to render their seeming eccentricities consistent forever.
>
> —Robert Walton, *Frankenstein*

The environments in *Frankenstein* seem to fall into one of two categories: sublime or domestic. In "*Frankenstein* with Kant: A Theory of Monstrosity

---

[13] This lyric's criticism of reified society is, in several senses, "immanent"; it speaks from a space of the interconnectedness of what humanism calls "nature" and "culture." Again, in "Cultural Criticism and Society," Adorno writes, "Dialectics means intransigence towards all reification. The transcendent method [of critique], which aims at totality, *seems* more radical than the immanent method, which presupposes the questionable whole. ... The choice of a standpoint outside the sway of existing society is as fictitious as only the construction of abstract utopias can be. Hence, the transcendent criticism of culture, much like bourgeois cultural criticism, sees itself obliged to fall back upon the idea of 'naturalness,' which itself forms a central element of bourgeois ideology. *The transcendent attack on culture regularly speaks the language of false escape, that of the 'nature boy.'*" (31, italics mine). Yet, he follows this with a reminder, perhaps to himself as much as to the reader: "*No theory, not even that which is true, is safe from perversion into delusion once it has renounced a spontaneous relation to the object.* Dialectics must guard against this no less than enthrallment in the cultural object. It can subscribe neither to the cult of the mind nor to hatred of it. The dialectical critic of culture must both participate in culture and not participate. ... The traditional transcendent critique of ideology is obsolete. In principle, the method succumbs to the very reification which is its critical theme. *By transferring the notion of causality directly from the realm of physical nature to society, it falls back behind its own object*" (33, italics mine).

or the Monstrosity of Theory," Barbara Claire Freeman writes that the novel's sublime topography "produces neither peace of mind nor aesthetic pleasure, but rather a vision of and an encounter with monstrosity" (195). Freeman argues that the link between Victor's encounters with the monster and sublime landscapes undoes Kant's formulation of the sublime as an index of the human mind's superiority over nature. Not only does the novel undo this formulation, it reverses it. Historically, the Alps have been a source of tremendous fear and anxiety; as Jane Nardin notes, even at end of the seventeenth century the glaciers around Chamonix were still exorcised by bishops (442). Nardin discusses the history of the Alps as a symbolic and literal meeting place for science's battle against superstition, and mountaineering as a byproduct of science's victory. She argues that Shelley had Saussure and other scientific explorers in mind when she wrote *Frankenstein*; the meeting between Victor and the monster on the Mer de Glace graphically demonstrates that the Enlightenment project has gone awry: the landscape formerly believed to be populated with demons now actually harbors one, one which is a product of science rather than the Devil (446).

In the minds of certain characters, sublime landscapes contrast sharply with domesticated (or domestic) ones. Walton's wild Arctic emerges against an idea of England as a warm, green home, invoked by the textual presence of Mrs. Margaret Saville. Walton's married sister, like the idea of landscape she suggests, functions as the frame of the novel. As the intended recipient of Walton's correspondence, present only through her absence, Mrs. Saville stands in for the author as well as the audience (as William Veeder, Fred Botting, and others have noted, it is not a coincidence that Mary Wollstonecraft Shelley and Margaret Walton Saville have the same initials). This dual relationship of author and audience embodied by Mrs. Saville contrasts with the binary of wild or sublime landscapes and domesticated ones, which parallels Walton's, and Victor's, binary view of nature and culture.

For Victor, Elizabeth Lavenza embodies domesticated nature, a space of absence cultivated with human presence. "The world was to me a secret, which I desired to discover; to her it was a vacancy, which she sought to people with imaginations of her own" (21). Here Victor attributes to Elizabeth a quality of his own, denying in himself and in his work the relationship between knowledge and oppression, between aspects of the Enlightenment project and imperialism, patriarchy, and racism. It is Victor, not Elizabeth, who seeks to people the world. Victor's characterization of Elizabeth also suggests that, to some extent, Victor viewed her world as empty, rather than the other way around; Elizabeth never says anything herself (in Victor's retelling) that would support his view of her relationship to nature. In fact, the pieces of remembered conversation Victor narrates suggest otherwise, such as her opposition to the elder Frankenstein's plans for Ernest Frankenstein to become a lawyer. Elizabeth suggests that Ernest (the only member of the Frankenstein family to survive Victor's science project) should instead become a farmer, because "he is continually in the open air, climbing the hills, or rowing on the lake ... A farmer's life is a very healthy happy life; and the least hurtful, or rather the most beneficial profession of any" (45). As

Anne K. Mellor argues, Elizabeth's plan for Ernest demonstrates a commitment to an ethic of mutual dependence and cooperation ("Possessing" 284). Far from viewing the world as empty, Elizabeth sees the world as a presence. Unlike Victor, Elizabeth seems to desire harmony with nature rather than the pursuit of nature "to her hiding places" (36). For this reason, Victor's choice of the word "vacancy" in this context (and that of "Mont Blanc") is particularly telling.

There are few actual descriptions of nonhuman nature in the novel, aside from the characteristically sublime (and so philosophically charged) landscapes of the Arctic, Alps, and Orkneys. Exceptions include the following cursory description of Victor's recovery from the shock of seeing his creature animate: "I remember the first time that I became capable of observing outward objects with any kind of pleasure, I perceived that the fallen leaves had disappeared, and that the young buds were shooting forth from the trees that shaded my window. It was a divine spring …" (43–4). Passages in the novel relate the fact of but do not depict Victor's perambulations in the countryside of Ingolstadt and the family walks in Plainpalais; they do not actually describe flora, fauna, or topography. The above passage is one of the few in the novel in which "ordinary" nature is observed and described. From the time of Victor's journey home (upon receiving news of William Frankenstein's death), the novel resides primarily in sublime landscapes; even the monster's narrative of his escape from the laboratory into the forest does not convey a clear sense of surroundings. "It is with great difficulty that I remember the original area of my being: all the events of that period appear confused and indistinct" (79). The monster recalls his initial environment in terms of his varied, confused sensations, particularly of moonlight and birdsong. Observation of his senses gives way to the observation of natural laws, such as cause and effect (in the form of fire). These primarily form a record of an initiation into consciousness, from sensation to reasoning.

In all three narratives of *Frankenstein*, nature serves to mark or reflect the development of consciousness. For Robert and Victor, the natural world often seems either evil or empty, whore or virgin; both the sublime and the "ordinary" (domesticated) are feminized (in two separate articles, Mellor suggests both that the creature represents the sublime and presents "an image of uninhibited female sexuality").[14] They stand transfixed in awe of mountains, oceans, and ice, while domesticated nature suggests an emptiness that at once signifies purity and a lack of power. However, the creature stands in a different relation to the rest of the world; he does not sharply distinguish between sublime and "ordinary" nature. He quickly finds sustenance and shelter in the woods, easily leaps onto Alpine peaks, and evenly weathers the Arctic cold. The world in fact almost welcomes him, that is, save the world of human beings. His attempts at reconciliation, segregation, and revenge enact a different form of the binary of nature and culture.

---

[14]   See "*Frankenstein* and the Sublime" (102) and "Possessing Nature" (279).

Though a number of critics[15] view the monster as a representation (and, to varying degrees, a critique) of Jean-Jacques Rousseau's "natural man," it is not uncommon for critics to read the monster as a direct representation of nonhuman nature itself. Indeed, Mellor's argument, that the novel critiques "possessing" nature through a depiction of its resistance to and eventual revenge upon Victor,[16] figures the creature as nature writ large. In "Frankenstein and the Sublime" she concludes that the creature represents nature as the sublime (embodying sublime landscapes and the sublime itself through his size and transgressive identity), depicting the human struggle with an unknowable nature (101–2). Mary Poovey even goes as far as to suggest that the novel signifies Shelley's fundamental distrust of nature, human and otherwise (256–8).

Those critics interested in the problem of nature and culture tend to refer to the third narrator as the creature (or Creature) and view him as just such an embodiment of nature, while the many critics interested in monstrosity as such tend to refer to the third narrator as the monster and read him as a signification of the unnatural. Chris Baldick returns to the Latin root of monster, *monstrare*, to reveal that "a 'monster' is something or someone to be shown," a visible example of the results of vice (*In Frankenstein's Shadow* 10); he defines "monster" as "one who has so far transgressed the bounds of nature as to become a moral advertisement" and the monstrous as "any ridiculous or unnatural combination" ("The Politics" 50, 51). This definition reveals the monstrous as that which does not meet moral or aesthetic norms (which are, of course, categories based on ideas of nature). Peter Brooks suggests that, on one level, the monster is an idea of body embodied, "nothing but body: that which exists to be looked at, pointed to, and nothing more" (102).

In these differently oriented critical works, the third narrator (as creature or monster) functions more as symbol than subject. Poovey's reading of the novel is one of the few that suggests that the third narrator is both a monster and an embodiment of nature. She argues that the monster is made (hence its monstrosity), and so,

> the monster cannot have independent desires or influence its own destiny because, as the projection of Frankenstein's indulged desire and nature's essence, the creature is destiny. Moreover, because the monster's physical form literally embodies its essence, it cannot enter the human community it longs to join, and it cannot earn the sympathy it can all too vividly imagine. Paradoxically, the monster is the victim of both the symbolic and the literal. (258)

The monster's subjectivity often proves a problem for critics and characters alike.[17] James McLarren Caldwell notes that the disjunction between mind and

---

[15]   Such as Paul Cantor, David Marshall, Nancy Yousef, and Anne McWhir.

[16]   As a representative of patriarchal science and its rape of nature (282).

[17]   As Lawrence Lipking argues, the monster's complexity, and the novel as a whole, presents genuinely insoluble problems (318–19), problems that do not allow for the ease of symbolic readings. In fact, Barbara Johnson views the central transgression of the

body that is monstrousness causes some characters to project bestiality onto the monster because they cannot tolerate the disjunction (270). He argues that the novel serves as a "heuristic exercise"; by repeatedly dramatizing the failure of sympathy, readers learn to sympathize with the monster, with a sympathy based not on uniformity or identification but on the accommodation of difference (270–71).[18]

Difference challenges rationalism and the kind of science based on its reductive view of the world.[19] Shelley's depiction of Victor as a scientist is, as Ludmilla Jordana has pointed out, a critique of nineteenth-century science's attempt to create a secure, disinterested cultural identity for itself as a controlled, masculine "unveiler of nature" rather than an unleasher of the monstrous (72–3). Marilyn Butler's *The Shelleys and Radical Science* locates the scientific context of the novel in the William Lawrence/John Abernathy Vitalist Debate:

> Frankenstein the blundering experimenter, still working with superseded notions, shadows the intellectual position of Abernathy, who proposes that the superadded life-element is *analogous to electricity*. Lawrence's skeptical commentary on that position finds its echo in Mary Shelley's equally detached, serio-comic representation … (xxi, italics mine)

The skepticism Butler refers to is inherent in Lawrence's empirical methodology and his corresponding opposition to Abernathy's extra-sensory conclusion; based on observation, Lawrence argued that life is intrinsic to living things, that "the motion proper to living bodies, or in one word, life, has its origin in that of their parents" (*Introduction* 142). Butler also suggests that the monster, like Lawrence, is a good, empirical scientist, taking on the task that Victor abandons: the scientific observation and documentation of Victor's technological achievement ("*Frankenstein*" 307–8). As a materialist critique of reductive science, the novel situates the monster as both the object of study and the student, suggesting the categories of "nature" and "culture" are ethically and scientifically untenable.

The Production of Monstrosity

Before moving into a dialectical approach to ecocritical concerns in the text, it would be useful to consider some of the many Marxist analyses of the novel. A number focus on a combination of the sociopolitical context of the novel and the relationship between Victor and the monster. For Paul O'Flinn, for example, the

---

novel as this very symbolic, or autobiographical, desire: Victor's desire to reproduce himself (243, 248).

[18] While David Marshall suggests that *Frankenstein* demonstrates an ambivalent view of sympathy, critiquing it as, in part, incestuous (212–13).

[19] Many critics, such as Crosbie Smith, read *Frankenstein* as a subversion of Enlightenment ideology, while Michie reads the novel as a subversion of that aspect of Romantic ideology that participates in the problematic assumptions of the Enlightenment.

Luddite disturbances of 1811–1817 are particularly significant.[20] Warren Montag observes that the French Revolution is (unlike the Luddite disturbances) alluded to in the novel but only obliquely; he dates the setting of the novel to the 1790s, in the midst of the French Revolution, through Victor's brief mention of the English Revolution of 1642: "It is indeed remarkable that the work refers to a revolution that occurred 'more than a century and a half before' rather than to the most important event of contemporary history" (385).

The textual absence of workers and the French Revolution has led many critics to read the monster as the incarnation of either or both. For example, while Baldick views the monster as a representation of Shelley's ambivalent feelings about the Revolution (61), some critics complicate this symbolic correspondence of the monster and the proletariat or revolutionary force. Montag argues that both Victor and the monster are products of capitalist reason, of the central horror of the novel: scientific technology, present only in its absence (391–2). Again, the modern world is equally absent from the text: factories and mills, workers and work, the urban and industrial. For Montag, this leads to the conclusion that the monster is, finally, not simply a representation of the working class but of its absence, the suppressed modern world and the unrepresentability of the proletariat (395). Created before the notion of a postcapitalist world, Montag argues that the novel can only "turn backward toward a time of mutual (if unequal) obligation, to a time before the creation of monsters by the industrial order, a time when the human was regulated by the natural" (395).

And herein lies the difficulty, from an ecocritical point of view, with most Marxist readings of *Frankenstein*: they tend to view the representation of nonhuman nature in the novel as either "part of the problem," repressing the modern (Montag 395), or as padding, filler from the Shelleys' "tourist diary," *History of A Six Weeks' Tour* (O'Flinn 26). Most do not see the way in which representations of nature connect to the core contradictions of the novel (its representation or lack of representation of class politics) and the way in which they are themselves inherently political. Elsie B. Michie's "*Frankenstein* and Marx's Theories of Alienated Labor," however, comes closest to this realization. While noting that a simplistic symbolic reading of the text situates the monster

---

[20]   These mass actions against machines in England continued through 1817, coincident with the Pentridge rising of June 1817, in which 300 men attempted to overthrow the government in a march toward Nottingham, expecting similar marches across the country; the group was captured and three of the groups' leaders were executed. O'Flinn writes, "Mary and Percy returned to England from Geneva in September 1816 and Luddites were still being hanged in April 1817 as Mary made the last revisions to her manuscript. Before *Frankenstein*'s publication in March 1818, [Percy] Shelley reacted to the execution of the leaders of the Pentridge rising with *An Address to the People on the Death of Princess Charlotte*, a forceful political pamphlet published in November 1817 and eagerly read by Mary, as she noted in her journal. The pamphlet lamented the 'national calamity' of a country torn between abortive revolt and despotic revenge—'the alternatives of anarchy and oppression'" (25).

as a representation of the proletariat, she argues that using Marx's descriptions of alienation to read the novel closely reveals Victor as the alienated worker and the creature as the externalization of his alienation (94–5). From this perspective, the materials used to make the creature signify production's breakdown of the natural world into the dead components of manufacture, as well as the worker's alienation from the natural world, his senses, and the materiality of production (the monster figures here also as a representation of materiality, signified by his incredible size) (96). While Michie doesn't spend much time examining depictions of nonhuman nature or the environment, she does acknowledge that "nature" might actually serve a purpose in the text beyond a repression of the modern.

Monstrosity and the Dialectics of Negation

Frankenstein moves like an iceberg in chill waters. Structurally, the narrative is surrounded by the Arctic Ocean and, within each concentric narration, a body of water serves as the background for the novel's most dramatic action. The move from freezing to temperate waters and back again parallels pre-birth, life, and death. Like an amniotic sac of creation, either Lake Leman, the North Sea, or the Arctic surround the meetings between Victor and the monster. It is particularly apt that Freud's correspondent termed this innate sense of unbounded nature—of unbounded connection with the universe—"oceanic." The word signifies origin, embodiment, and embeddedness, literally and metaphorically.

The narrative structure of *Frankenstein* embeds human activity in, quite literally, a sea of connections. Bodies of water seem present in opposition to other bodies: mountains, islands, and "man." Here expanses of water function as sublime landscapes in their own right, as unfathomable immensities and impenetrable depths. And, like typically sublime topography, these large bodies of water have peaks and valleys, jagged edges and roaring sounds; they are uncontrollable and overwhelming. As emblems of nonhuman nature, they *seem* to stand in opposition to culture and everything human.

Far from functioning as a retreat from the problems of modernity, there is a connection between these *natural* bodies of water and the *produced* body of the monster. Victor's drive to transcend human nature and culture, to transcend the limitations of the human body and human knowledge, is akin to Robert's drive to transverse the Arctic Ocean. In both cases, materiality becomes an obstacle to overcome, rather than the fabric of existence. The constant narrative proximity to and within bodies of water, and its parallel in the body of the monster, serves to remind us of *our* bodies, and the human place in the material weight of the world.

Again, just as the lyric poem is often the most social when it seems the least so, positive and negative inscriptions may occupy the same textual space. What is negatively inscribed in *Frankenstein* is what the text means but cannot always understand or say. Similarly, the monster or the monstrous is that which speaks but isn't allowed to "mean." The monster is that which speaks or acts willfully, thoughtfully, and *is not human*; the monster, the ultimate Other in humanist culture, serves to deny the existence of this worldly nonhuman agency, to relegate it to the shadows and the fantastic. While the monster signifies many things (including, as we have seen, the unsignifiable itself) he is also, negatively, an imprint of the human fear of nonhuman agency.

The pervasive reading of the monster as symbol (for nature or this or that) elides his radical agency, the very essence of his textual life. As the Bildungsroman of Frankenstein's monster (and of Frankenstein himself) the novel speaks to our relationship with our origins and the origin of culture itself as a response to nonhuman nature. The monster's narrative, embedded within Victor's story told to Robert, details his birth to sensation, thought, and reason. As he watches the De Lacey family in their cottage from the hole in the wall of his adjacent refuge, the monster learns about human beings and human culture through observation and second-hand lessons. In a remarkably short period of time, the monster leaps from the realization that the sounds he hears have meanings to the association of words with objects:

> I found that these people possessed a method of communicating their experience and feelings to one another by articulate sounds ... This was indeed a godlike science, and I ardently desired to become acquainted with it. But I was baffled in every attempt I made for this purpose ... By great application, however, and after having remained during the space of several revolutions of the moon in my hovel, I discovered the names that were given to some of the most familiar objects of discourse: I learned and applied the words fire, milk, bread, and wood. I also learned the names of the cottagers themselves. (88–9)

Within another few months, the monster has already grasped the idea of reading and writing:

> This reading had puzzled me at first; but, by degrees, I discovered that he [Felix] uttered many of the same sounds when he read as when he talked. I conjectured, therefore, that he found signs on the paper for speech which he understood, and I ardently longed to comprehend these also; but how was that possible, when I did not even understand the sounds for which they stood as signs? (90)

The monster's narrative of his development does not make any attempt to explain how it was possible for him to grasp such concepts so quickly, though he masters language itself in a matter of a few more months. As Nancy Yousef argues, the monster's development is in most respects no less fantastic than his non-birth (198).[21]

---

[21] Yousef situates the impossibility of this development in the context of a sustained, multi-generational feminist engagement with empiricist thinkers such as Locke and Rousseau; the monster's radical autonomy, from the dependency of infancy and the normal childhood network of relationship (which, she argues, serves to critique and enlarge the tradition of empiricism) makes him monstrous (220–26).

The monster's speech, his narration to Victor (reported by Victor to Robert, narrated by Robert in a letter to his sister Margaret [Walton] Saville), is quite literally the center of the novel, and his acquisition of language and knowledge the most improbable part of his existence, the center of his impossible existence. His voice is the voice of question, the voice of the child discovering the world and its misery as his own. Upon learning of the nature of the human social world from Felix's recitations to Safie from Volney's *Ruins of Empires*, the monster begins to examine his own nature:

> The words induced me to turn towards myself. I learned that the possessions most esteemed by your fellow-creatures were high and unsullied descent united with riches. A man might be respected with only one of these acquisitions; but without either he was considered, except in very rare instances, as a vagabond and a slave, doomed to waste his powers for the profit of the chosen few. *And what was I?* Of my creation and creator I was absolutely ignorant; but I knew that I possessed no money, no friends, no kind of property. I was, besides, endowed with a figure hideously deformed and loathsome; I was not even the same nature as man … When I looked around I saw and heard of none like me. Was I then a monster, a blot upon the earth, from which all men fled, and whom all men disowned? (96, italics mine)

Moments later, the monster self-consciously repeats his question, "*What was I?* The question again recurred, to be answered only with groans" (97, italics mine). Even before he learns of his fantastic creation at the hands of Victor, the monster is convinced that he is wholly different from human beings, "a blot upon the earth." Only after he learns of the hierarchical organization of Western culture does he asks, "What was I?" It is the mirror of the question of the final lines of "Mont Blanc," "And what were thou, and earth, and stars, and sea,/ If to the human mind's imaginings/ Silence and solitude were vacancy?"

The monster asks his terrified question, which positions him as a monster even in relation to himself, not when he discovered the horror of how he was made in Victor's laboratory, but when he discovers *from what* and *into what* he has been made: the past and present of Western culture. Read positively, the horror that the monster expresses is a horror of culture; it, and not the monster, is in fact the subject of the monster's query. Read negatively, this question becomes the sound of the subject, of culture itself in the figure of the monster, asking monstrosity, "What are you?" Displaced grammatically as object, the referent of "what" rather than who, the voice of Western culture asks, through its own subject/product, the nature of its "opposite." The monster is, in this sense, that which literally figures the horror of nonhuman agency. While the content of the question seems designed to deny the agency of the nonhuman, as a question it formally acknowledges the agency (indeed, even the subjectivity) of nonhuman others. It is a question to which humanist culture fears, but does not expect, an answer. It is the question of the child who asks, quietly, in the dark, "Is there anyone under my bed?"

Language, defined as a system of communication that signifies abstract concepts as well as concrete referents, is supposedly that which separates human beings from other creatures and, correspondingly, culture from nature. Language represents the difference between human and nonhuman as the difference between agency and nonagency. In "Teaching the Monster to Read: Mary Shelley, Education and *Frankenstein*," Anne McWhir notes that the eighteenth century contained many beings whose status as human was controversial, including wild men, idiots, women, non-Western peoples, and orangutans: "According to James Burnet, Lord Monboddo (1.187–88), [orangutans] were in fact members of the human species. Monboddo argues this on the basis of their educability and their presumed capacity for speech" (80).[22] Western culture's narrowly defined notion of language serves as the benchmark for subjectivity, ordinarily to exclude, rather than include, other living beings in the realm of meaningful consciousness and therefore agency and rights. Certainly the monster's very narrative challenges the notion that language, and therefore culture, is the sole territory of human beings; he and others repeatedly emphasize both his difference from human beings and his remarkable powers of articulate reasoning. To share our identity as culture-makers with other beings means, terrifyingly, that we share their status as part of and subject to nature, "red in tooth and claw" or otherwise. The degree to which Shelley's depiction of the monster's process of acquiring language—Rousseau's "supplement" to nature—seems fantastic is the degree to which we fear nonhuman agency.

Explicitly, the monster's ability to narrate makes him monstrous, makes him "unnatural." Negatively, his narrative of coming into culture expresses humanist culture's anxiety about its self-proclaimed identity as unique in the world. A little nervously, culture exclaims, "I am alone; everywhere I turn I see that there is nothing else like me in the world." In this way, the monster functions as an expression of our fear, whispering against our insistence that we are alone. The monster's lack of "natural" connections, his manufactured alienation, negatively betrays our recognition, and fear, of similarity. The novel repeatedly compares the monster to other characters, encouraging readers to do the same, to recognize similar needs, desires, and fears. As nearly all critics of the novel note, the monster is, to some degree, Victor's double.

This representation of duality suggests a conscious critique of capitalism as well as an anxiety about identity within capitalism. Margo Perkins's Hegelian reading of the novel focuses on the Master-Servant dialectic between Victor and the monster and its relation to class politics, especially in the representations of the relationships between class and justice, and material reality and ethical values

---

[22]   In the notes to the *Discourse on Inequality*, Rousseau writes, "Without ceremony our travelers take for beasts, under the name *pongos, mandrills, orangutans*, the same beings that the ancients, under the names *satyrs, fauns, sylvans*, took for divinities. Perhaps, after more precise research, it will be found that they are neither animals nor gods, but men" (McWhir 209).

(27). Perkins notices that Victor imposes the moral values of his class onto the monster and that, according to those values, characters' responses to the monster are predetermined; as we see with little William Frankenstein, there can be no innocence within capitalism (39–40). Perkins argues that although Shelley raises issues about class oppression, she doesn't really challenge the existing class structure (40). Indeed, as a representation of class relations, the dialectic between Victor and the monster does not seem to allow for the possibility of positive change, for an intervention in the spiraling crises of capitalism. From this perspective, the novel reifies bourgeois anxiety about class oppression, making it another regrettable yet insurmountable "natural" crisis, like a storm above Mont Blanc or turbulence on Lake Leman. It becomes, as a product of a Victor's "natural" curiosity and human failings, part of the cost of living.

However, as a double (and to some degree parallel) Bildungsroman, *Frankenstein* expresses this anxiety in more fluid terms. Victor's narrative of coming into culture begins with the social context of his birth, the social standing of his family in Geneva, and the history of his parents' meeting and marriage. The first mention of Victor's person is in the context of his education: "[My father] relinquished many of his public employments, and devoted himself exclusively to the education of his children. Of these I was the eldest, and the destined successor to all his labors and utility. No creature could have more tender parents than mine" (19). From this beginning, Victor goes on to describe the introduction of his cousin Elizabeth into his family, his close friend Henry Clerval, his interest in natural philosophy, and the violent thunderstorm that inspired his later experiments. At seventeen, "I became the instructor of my brothers. Ernest was six years younger than myself, and was my principal pupil" (25). It is significant that Ernest (again, the only member of the Frankenstein family to survive the events of the narrative) is Victor's primary student, a point I will return to in a moment.

Victor's acculturation is, with the exception of his father's rebuff concerning Cornelius Agrippa, a happy discovery of the world, as a student and teacher. The monster's narrative, on the other hand, begins with pain and uncertainty about the world and his place in it, and is followed by his realization that he is unwanted, even by his creator—that he does not have a place in the world at all. The contrast between one creature's fond description of his tender parents and the other's sense of his parent's horror at the fact of his existence is striking.

As a pedagogical tool, the Bildungsroman often serves as either cautionary or exemplary tale. That both Victor's and the monster's narratives revolve around education seems particularly important. Victor is not only Ernest's instructor, he is also the monster's. Through Victor's journal, the monster learns not only the facts of his birth but also the social context of his production. While Victor begins his narrative with the knowledge of his birth (natural and social), the monster must discover this information through writing. As Paul Cantor and others argue, the monster's acculturation only furthers his alienation (126), while human solidarity with Victor persists throughout the novel (Yousef 223). In fact, McWhir goes so far as to argue that the monster is an uncritical reader whose encounters with *Paradise Lost* and other texts cause only suffering.

The role of production in acculturation is expressed in the intertextuality of the parallel Bildungsromane. While the positive representation of this produced nature of culture betrays an anxiety about identity within capitalism, the negative of this representation reveals capitalist culture's own anxiety about its nature. While these Bildungsromane seem as different as two accounts of growing into the world could possibly be, they both clearly represent acculturation as a process of production, a process interpersonal and intertextual. The monster's narrative is embedded within Victor's, which is embedded within Walton's letters to his sister (breaking the traditionally male cycle of the genre as a story of male acculturation to be read by young men in the process of their acculturation).[23] Through the embedded external narrative of *Paradise Lost,* the novel alternately casts Victor as God-like, Adamic, and Satanic, and the monster as Adamic and Satanic, and each character periodically refers to himself in such terms. This revelation of the instability of produced identity, even between the extremes of creation itself, is the negative imprint of culture's anxiety about its own identity, its relation to nature and the nature of nature itself. Culture is produced, as the novel insists through its layers of intertextual Bildungsromane, and Victor and the monster are, in turn, products of acculturation and reproducers of culture. Negatively, this pattern reveals a troubling question, the same question imprinted in the monster's articulation, "What was I?": if nonhuman nature is as much "produced" as human nature, as evidenced by the monster, then what is human culture but one of the many cultures in nature?

As a pedagogical tool, one may wonder what this double Bildungsroman teaches and to whom. Ernest, as the only surviving member of the immediate Frankenstein family, is in some sense heir to this knowledge, though it is quite literally addressed to someone else. Perhaps, as Elizabeth had hoped, he broke with his father's wishes, and Victor's teachings, and became a farmer. While there is no evidence that Walton benefits from the lesson of Victor's experience, there is the possibility of Ernest, the possibility of a different relationship between "nature" and "culture," a different outcome of acculturation. As it stands, the processes of acculturation represented in the novel, the interconnected Bildungsromane, reflect the monster's horrified articulation, "What was I?": human culture's fear of nonhuman agency and the objectification of humanity (that is, human animality).

This fear, imprinted in the monster's narrative of his acquisition of language, is equally, and perhaps most clearly, present in Victor's initial reaction to the monster's gaze.

---

[23] With this pattern in mind, the monster's tale as the core narrative is the most intertextual, even to the point of invoking another Bildungsroman as part of the monster's acculturation, Goethe's *Sorrows of Young Werther.*

It was already one in the morning; the rain pattered dismally against the panes, and my candle was nearly burnt out, when, by the glimmer of the half-extinguished light, I saw the dull yellow eye of the creature open; it breathed hard, and a convulsive motion agitated its limbs.

How can I describe my emotions at this catastrophe, or how delineate the wretch whom with such infinite pains and care I had endeavored to form? His limbs were in proportion, and I had selected his features as beautiful. Beautiful!— Great God! His yellow skin scarcely covered the work of muscles and arteries beneath; his hair was of a lustrous black, and flowing; his teeth of a pearly whiteness; but these luxuriances only formed a more horrid contrast with his watery eyes, that seemed almost of the same color as the dun white sockets in which they were set, his shriveled complexion, and straight black lips ... Unable to endure the aspect of the being I had created, I rushed out of the room, and continued a long time traversing my bedchamber, unable to compose my mind to sleep. (38–9)

Victor's shock at the sight of the monster's eyes is truly shock at the sight of the monster looking at him. It is not the aspect of his eyes, their color, or the composition of the monster's visage (which, presumably, Victor was already quite aware of) but their aspect imbued with life, the monster's gaze, that causes Victor to flee in "breathless horror and disgust" (39). Negatively, this fear is the fear of objectification, the fear of being on the other end of subjectivity. To the mind of humanist culture, the very fact of the nonhuman other's subjectivity, embodied in his gaze, threatens to do what (human) subjectivity—what human culture—often does: objectify and dominate.

Victor fears being produced as an object in his object's gaze; in rationalist culture, the spectator is always superior to the spectacle, the subject to all he senses. Negatively, this is the nightmare knowledge of Western consciousness, its life as the passive, helpless object of nature's agency. Just as it has made agential nonhuman nature monstrous, it fears being made monstrous. Sight, like language, represents subjectivity, which really boils down to power. Here vision serves as a mark of consciousness, and so nonhuman nature's vision of humanity (the monster's gaze at Victor) threatens this construction of human agency. Humanist culture goes to great lengths to insist that nonhuman animals look but do not see, make sound but do not speak. Victor's fear of the monster's gaze is the negative imprint of culture's fear of objectification, a fear imprinted as early as the narrative of Jehovah's insistence that Moses look not upon his face for fear of death. Here too, power does not want to be seen, to be objectified in the gaze of another.

What is the relationship between the story of culture's reproduction, the Bildungsroman, and culture's fear of objectification, of oppression? The labyrinthine origins of culture, its production and reproduction, lead back to the phenomenon of the oceanic feeling and the discourse of monstrosity. The Bildungsroman, or story of coming into culture, is also the story of culture coming into us; its production is its reproduction ad infinitum. In this way, culture constructs itself as a "second nature," replacing "nature" as world, displacing it to the shadowy status of origin.[24] It is no coincidence that narratives of coming into culture always begin inside it; for humanist culture, there is no arrival from without, only the realization that we are always already inside this "second nature." From this point of view, the humanist genre of the Bildungsroman appears as an insistence that there isn't anything out there; no there *there*, only the here of culture. The oceanic feeling is the reminder that there is something, even someone, out there, but in the world of "second nature," this feeling is displaced, bifurcated into the sublime and the monstrous, the transcendent and the unnatural.

In *Frankenstein*, mountains and bodies of water appear by turns threatening or inspiring. The monster's relationship to the Alps characterizes them as monstrous, unnaturally large, jagged, and overwhelming, whereas Victor's experience of the mountains, before his encounter with the monster, characterizes them as the repository of an awesome transcendence. There is a similar dynamic with the many bodies of water in the novel, from Lake Leman to the North Sea and the Arctic Ocean. However, in the case of water, it is Walton, and not Victor, who frames the reader's experience, painting the Arctic as sublime. Here Robert describes Victor's reaction to the Arctic landscape: "Even broken spirited as he is, no one can feel more deeply than he does the beauties of nature. The starry sky, the sea, and every sight afforded by these wonderful regions, seems still to have the power of elevating his soul from earth" (16). Though perhaps more of a description of Robert himself than his mysterious passenger, contrast the previous passage with Victor's experience of the Orkneys and the North Sea:

> It was a place fitted for such work, being hardly more than a rock, whose high sides were continually beaten upon by the waves ... [W]hen the weather permitted, I walked on the stony beach of the sea, to listen to the waves as they roared, and dashed at my feet. It was a monotonous, yet ever-changing scene. I thought of Switzerland; it was far different from this desolate and appalling

---

24   The concept of "second nature" dictates that we cannot know or truly experience nonhuman nature because of our historical condition of alienation; if we could really experience nature we would not know it as such (only in alienation do we have the conceptual binary nature/culture). In *Marxism and Totality: The Adventures of a Concept from Lukács to Habermas*, Martin Jay discusses Georg Lukács's view of reification and second nature: "This term, one not in fact found in Marx himself, meant the petrification of living processes into dead things, which appeared as an alien 'second nature'" (109). While Lukács explicates second nature as a product of reification, Murray Bookchin's school of social ecology views second nature as part of an evolutionary teleology that reaches fruition with humanity as the world's consciousness: "Second nature is, in fact, an unfinished, indeed inadequate, development of evolution as a whole. ... Humanity as it now exists is not nature rendered self-conscious. ... Free nature [a sort of third nature], in effect, would be a conscious and ethical nature, embodied in an ecological society" (www.greeninformation.com/socialecology.htm). Sadly, Bookchin does not seem to include other species as citizens in this "ecological society." The concept of second nature allows him to continue to construe human and nonhuman natures in the old, hierarchical terms: nonhuman nature as matter, human nature as mind.

landscape ... Its fair lakes reflect a blue and gentle sky; and, when troubled by
the winds, their tumult is but the play of a lively infant, when compared to the
roarings of the giant ocean. (136–7)

Later, after Victor's destruction of the half-formed female creature, he revisits the
same beach, which he "almost regarded as an insuperable barrier between me and
my fellow creatures; nay, a wish that such should prove the fact stole across me"
(141). The size and sound of the ocean, compared to the infant-like size and swell
of Lake Leman, places it in the company of the unnaturally large monster and
the Alps that he leaps across with agility. Described as beating and roaring, the
ocean's violence seems both typically oceanic and oddly simian.

    The Arctic Ocean, Robert Walton's "country of eternal light" (5–6) is also the
landscape into which the monster disappears, "borne away by the waves, and lost
in darkness and distance" (191). These contrasting views of the same landscape,
and the seeming duality of representation of the environment in general, as
either transcendent/sublime or unnatural/monstrous, reflect the environmental
conditions of the novel's production.[25] Again, both *Frankenstein* and *History of
a Six Weeks' Tour* were composed, in greater part, during the coldest, wettest
summer in Europe in a hundred years. As Clubbe writes, the weather of 1816,
during which Lake Leman flooded, "may even be the single most determining
influence upon the novel's creation" (27); he links a moment in *Frankenstein*,
quoted below, to a real moment in Shelley's experience. As noted earlier in this
chapter, Victor sees the storm "from precisely the spot in which on 10 June Mary
Shelley had originally seen it surging across the waters and had described it in
much the same words" (34).

    I quitted my seat, and walked on, although the darkness and storm increased
every minute, and the thunder burst with a terrific crash over my head. It was
echoed from Salêve, the Juras, and the Alps of Savoy; vivid flashes of lightning
dazzled my eyes, illuminating the lake, making it appear like a vast sheet of
fire; then for an instant every thing seemed of a pitchy darkness, until the eye
recovered itself from the preceding flash. ...

    While I watched the storm, so beautiful yet terrific, I wandered on with a
hasty step. This noble war in the sky elevated my spirits; I clasped my hands,
and exclaimed aloud, "William, dear angel! this is thy funeral, this thy dirge!"
As I said these words I perceived in the gloom a figure which stole from behind a
clump of trees near me; I stood fixed, gazing intently: I could not be mistaken. A
flash of lightning illuminated the object, and discovered its shape plainly to me;
its gigantic stature, and the deformity of its aspect, more hideous than belongs
to humanity, instantly informed me that it was the wretch, the filthy daemon to
whom I had given life. ... Nothing in human shape could have destroyed that
fair child. *He* was the murderer. (56)

---

    [25] As several critics have noted, the fantastic thunderstorms that move throughout the
text, and indeed all of the descriptions of the Alps in the novel, are drawn from Shelley's
actual time in the region recorded in *History* and in her letters.

Clubbe attributes the role of lightning in the novel, which appears as a creative and destructive force at key moments in the text, to the incredible storms the Shelleys witnessed that summer. His work suggests that it isn't a coincidence that *Frankenstein* is a tale "of the human psyche in agonized conflict with the supernatural" (26).

Again, the weather, indeed the entire environmental conditions of the novel's composition, imbued *Frankenstein* with culture's terror of nature's agency—the agency of oceans and storms and jagged peaks—as the horror of the nonhuman in the figure of the monster. The conflict with the supernatural is, in fact, a conflict with the idea of being part of nature, with the fact of human animality. The above passage is significant not only because it links Shelley's lived experience of the nonhuman world to the text, but also because it links the power and agency of nonhuman nature to the person of the monster (this is reflected in form as well as content; just as the ocean seems somewhat simian in a later passage, here the monster appears almost elemental). As these two are fused, so are Victor and the monster, making Victor the murderer, rather than the keeper, of his brother. Victor's moment of "elevated" spirits, in which his solitary grief and rage seem passively reflected in the environmental conditions around him, becomes one of horror and disgust with the appearance of the monster. In this same scene, Victor refers to him as wretch, daemon, devil, being, creature, and, significantly, *"the animal"* (57, italics mine). Though he considers the monster "nearly in the light of my own vampire, my own spirit let loose from the grave, and forced to destroy all that was dear to me," Victor denies his own animality when he calls the monster "the animal" (57). Again, "nothing in human shape" could have destroyed the "angel" William, only something animal or elemental, "a blot upon the earth," as the monster calls himself, "a disfiguring spot or mark" ("Blot"). Uncannily, the monster is also both "a moral stain" and "an obliteration by way of correction" ("Blot"); he embodies death and destruction both as an example of and corrective to the arrogance of humanist culture. In several senses, he is also a reminder of culture's status as product of the natural world. The supernatural language (the "angel" William and "vampire" monster) and plot of the novel are displacements of (and placeholders for) a far more earthly and earthy terror.

An ecocultural reading of *Frankenstein* suggests that the influences of nonhuman nature permeate content and form. The connections between nonhuman nature and the produced body of the monster surface in the monster himself—a nonhuman creature and a product of human culture (and acculturation). *Frankenstein* registers the terror of nonhuman agency as, in part, the humanist anxiety of objectification. In "second nature" the oceanic feeling becomes the fear of monstrosity. The horror of *Frankenstein* is a horror of ourselves.

Chapter 4
Placing Modernity in *Orlando*

Modernism is arguably both an heir to the project of the Enlightenment and a revolt against its historical process. This ambivalence is variously manifest in the presentation of the modern "subject." Modernism cannot really make the "loss" of the bounded bourgeois subject and the breakdown of its values a part of its discourse without in the first place invoking the validity, however tentative, of that subject and those values ... Modernism thus invokes the bourgeois subject, but it does so more through negation than affirmation.

—Astradur Eysteinsson, *The Concept of Modernism*

The wanderer, the loner, the exile, the restless and rootless and homeless individual were no longer the rejects of a self-confident society but rather those who, because they stood outside, were uniquely placed in an age when subjectivity was truth to speak with vision and authority.

—James McFarlane, "The Mind of Modernism"

"Placed" in the Alps, both *Candide* and *Frankenstein* are fantastic narratives of wandering. Indeed, Candide's catastrophes and the monster's existence require a suspension of disbelief of the sort that travelers to the Alps in the eighteenth and even nineteenth centuries claimed necessary in the face of those "horrid crags" and eternal peaks. Set in the contrasting climes of England and Constantinople, Virginia Woolf's *Orlando* presents us with a character whose wandering through four centuries of experience also defies human logic and the "laws of nature." Just as Romanticism continues or is in conversation with the Enlightenment, Modernism shares many of the interests and concerns of Romanticism.[1]

The characteristic often used to define Modernism against earlier movements has been its cosmopolitanism or, in contrast specifically to Romanticism, its seeming placelessness. For this reason Carol H. Cantrell writes that modern art "would seem to be a hostile territory for a student of literature and the natural environment" (25). And yet, as Cantrell argues, modernists witnessed profound changes in the planet, "a revolutionary change in the 'the given,' including 'the given' we call nature" as a product of human culture and, as a result, implicit "in the modernist aesthetic project is a critique of [the] Western understanding of reason ... based on the separation of perceiving mind from the perceived world"

---

[1]   N. Takei da Silva argues, "In respective roles as bequeathor and instigator of the Modernist principles in art and literature, Romanticism and Freudian psychology may be considered the two spiritual roots of the Modernist Movement" (ii). Woolf's explicit interest in both psychological ideas and Romanticism, he continues, "her attempt to revive its [Romanticism's] essential spirit," crowns her "a most fitting representative or the embodiment of the spirit of Modernism" (iv).

(26). Cantrell goes on to note that Modernism and ecological literary criticism are both concerned with representations of otherness and the problem of language and place that play central roles in the modernist articulation of alienation and fragmentation of experience.[2] Indeed, the lines of connection between Modernism and ecocriticism as critiques of modernity are complex and manifold.

The supposed "placelessness" of Modernism is itself a placeholder for anxiety about modernity, in all its fast-paced fragmentation and urban confusion.[3] This critical anxiety surfaces in part from Modernism's characterization as either deserter or collaborator in the crisis of modernity. Astradur Eysteinsson's *The Concept of Modernism* argues that Modernism has been consistently confused with an "escape from history" (viewed through the principles of New Criticism) or reduced to a mirror of social modernization. Instead, Modernism should be viewed as an intervention or, as Eysteinsson phrases it, an interruption:

> [A]n attempt to interrupt the modernity that we live and understand as a social, if not "normal," way of life. Such norms are not least buttressed by the various channels and media of communication and this is where the interruptive practices of modernism appear in their most significant and characteristic forms. In refusing to communicate according to established socio-semiotic contracts, they seem to imply that there are other modes of communication to be looked for, or even some other modernity to be created. (6–7)

In a reified culture, such art resists society in its very existence outside efficient modes of communication. Quoting Adorno, Eysteinsson writes, "The asocial aspect of art 'is the determinate negation of a determinate society'" (41). Hence, the irrationality of Modernism is a mediated mimesis:

> In *Äesthetische Theorie* Adorno notes that the fact that mimesis is practicable in the midst of rationality, employing its means, manifests a response to the base irrationality of the rational world and its means of control. For the purpose of rationality, of the quintessential means of regulating nature, 'would have to be something other than a means, hence a non-rational quality. Capitalist society hides and disavows precisely this irrationality, whereas art does not.' Art holds forth the image, rejected by rationality, of its purpose and exposes its other, its irrationality. (42)

---

²   See Joanna Tapp Pierce's "Placing Modernism: The Fictional Ecologies of Virginia Woolf, Winifred Holtby, and Elizabeth Bowen" and Josephine Donovan's "Ecofeminist Literary Criticism: Reading the Orange" in *Ecofeminist Literary Criticism*. Ed. Greta Gaard and Patrick D. Murphy. Chicago: U of Illinois P, 1998, 74–96.

³   Unlike Romantic locodescriptive poetry and Alpine narratives, Modernism's relationship to place has often been represented as exclusively urban and semi-nomadic: the Paris-Berlin-London-New York nexus. Yet, as Malcolm Bradbury notes, Romanticism was also "an international movement of revolutionary sensibility; and it certainly marks the beginning of the aesthetic transition into the modern age" (*Social Context* 75).

This non-rational quality, the "end" of reason, cannot be anything other than the transcendence of the world. While Eysteinsson doesn't follow Adorno into the notion of art as *only* "unconscious historiography," he argues that a dialectics of Modernism links "artistic autonomy to a dialectical social mimesis," thus demonstrating Trilling's notion of Modernism's "adversary culture" (46, 222). Modernism's seeming placelessness (or social and natural "autonomy") is in this way partly mimetic and partly negative. This quality of placelessness expresses modernity's anthropocentric negation of place (that is, of the world) as the locus of meaning and, negatively, expresses critical (and, more broadly, cultural) anxiety about the agency of the more-than-human world.

As a text in dialectical relation with modernity, *Orlando* challenges rationalism both explicitly and negatively. The fantastic movement of time and space around one equally fluid individual exposes the absurdity of rationalism and the assumption of human superiority at its core. The more-than-human world becomes central to the meaning of the novel, problematizing the humanist subjectivity depicted in Woolf's mock-biography.

Orlando

Woolf's 1928 novel chronicles the development of a sixteenth-century English lord through four centuries of history, during which time he travels as Ambassador to Constantinople,[4] becomes a woman, lives with the gypsies of Broussa, returns to England, enters into and rejects literary society, and—at last—becomes a celebrated poet. Orlando is sixteen when we meet *him* and thirty-something when we leave *her* in the "present" moment, "Thursday, the eleventh of October, Nineteen Hundred and Twenty-eight" (228)—the publication date of the first edition of the novel.

Orlando touches life directly in several places; Orlando's character and his ancestral home (of 365 rooms) are loosely based on Woolf's friend and lover Vita Sackville-West and her formidable family estate in Kent, Knole House (also called the "Calendar House").[5] Shakespeare, Addison, Pope, and Swift make cameo appearances. Written partly as a gesture of love for Vita; partly as

---

[4]　As noted in the Introduction, I will refer to İstanbul throughout the text as Constantinople, as Woolf does in *Orlando*. İstanbul became the official name of the city in 1930.

[5]　In "Fact and Fantasy in *Orlando*: Virginia Woolf's Manuscript Revisions," Charles G. Hoffmann notes, "On permanent display to the public at Knole House, now a National Trust House, the bound, single volume manuscript of *Orlando* was presented as a gift by Virginia Woolf to Victoria Sackville-West on December 6, 1928, and inscribed, 'Vita from Virginia.' Writing of Virginia Woolf and *Orlando* in *The Listener*, Victoria Sackville-West states, 'Such things as old families and great houses held a sort of Proustian fascination for her. Not only did she romanticise them—for at heart she was a great romantic—but they satisfied her acute sense of the continuity of history, English history in particular'" (435).

feminist exploration of the social constructions of gender; partly as a parody of biography, Elizabethan to Edwardian; and partly as a theory of fiction, *Orlando* moves beyond the male literary canon and Thomas Carlyle's formulation that "history is the story of great men" while challenging Freud's claim that anatomy is destiny in the figure of Orlando himself. As a treatise against patriarchal history and biological determinism, the novel parodies Romantic lyricism as an appropriative gesture toward human and nonhuman others;[6] *Orlando*'s narrative structure (which parallels the title character's psychological, physical, and intellectual development) functions largely as a passage between, and enfolding of, Romanticism and Modernism.

Woolf began *Orlando* as a respite from "serious" writing. Of the novel's genesis, Mark Hussey states that, from the beginning, "Woolf saw *Orlando* as part comic, part serious. When she finished the manuscript and showed it to Leonard Woolf (who thought it 'in some ways better than The Lighthouse'), she reflected that she had probably begun it 'as a joke & went on with it seriously'" (201). Woolf was by no means the only one to refer to her novel in such terms: Arnold Bennett called *Orlando* "a high-brow lark," Conrad Aiken quipped that Woolf "expanded a *jeu d'esprit* to the length of a novel," and J.C. Squier dismissed it as "a very pleasant trifle" (Hussey 204).

Partly a by-product of his formulation of the unconscious in *The Interpretation of Dreams*, Freud's 1905 *The Joke and its Relation to the Unconscious*[7] suggests that jokes (including absurdity, irony, caricature, and parody) function psychologically in a fashion similar to dreams. "Joke-work," like "dream-work," economizes the psychic expenditure of libido by giving voice to repressed desires through the unconscious mechanisms of transformation, condensation, and displacement (159). While innocuous jokes create pleasure as a distraction from and rebellion against the rule of rational thought, tendentious jokes express aggression that is sexual and generally social (135–51). Of course, jokes are always social; they are "Janus-like," speaking out of two mouths, to self and other, at once. Arguably, when Woolf referred to *Orlando* as "a joke," she may have had something serious—and not simply dismissive—in mind.[8]

---

[6]   See Kari Lokke's "*Orlando* and Incandescence: Virginia Woolf's Comic Sublime."

[7]   Published in English in New York in 1916 as *Wit and Its Relation to the Unconscious* and reprinted in London in 1917 and 1922 (see the editor's preface, *Jokes and Their Relation to the Unconscious* [New York and London: Norton, 1960]). Freud writes, "The Janus-like double-facedness of the joke, which secures its original gain of pleasure against the attacks of critical rationality, and the mechanism of fore-pleasure belong to the first tendency [the creation of the joke]; the further complication of technique by the requirements spelt out in this section ["The Motives for Jokes—the Joke as Social Process"] arises from taking the joke's third person [audience] into account. So the joke is essentially a double-dealing rogue who serves two masters at once" (148–9).

[8]   Woolf's own relationship to Freud and psychoanalysis is certainly Janus-like; she railed against it while all around her, family, friends, and colleagues were enthusiastically exploring and promoting the new science in Britain (Adrian Woolf trained as an analyst,

Like Freud, Woolf relished jokes. In 1910, she and her friends, including Anthony Buxton and Duncan Grant, disguised themselves with costume, blackened faces, and beards to impersonate the Abyssinian Emperor and his entourage aboard the H.M.S. *Dreadnought*, the flag-warship of the Royal Navy at that time.[9] Woolf and her circle were aware of humor's subversive cultural value; after all, when all of the Empire is dressing-up British (a key subject for her friend E.M. Forster), these imposters present a reverse image of this absurdity to imperial power.

When Woolf repeatedly refers to *Orlando* as a joke it is, I argue, with a sense of the importance of the "double-facedness" of jokes and of fantasy. As Freud writes: "Nothing distinguishes the joke more clearly from all other psychical formations than its two-sidedness and two-facedness, and from this aspect at least, in their emphasis on 'sense in nonsense,' our authorities [on humor] have come closest to an understanding of the nature of the joke" (167). Against our natural "pleasure in nonsense," rationality dictates what behavior and which thoughts are acceptable according to strict social codes that have their origin in psychic needs. This "unnatural" rule of reason has a life of its own beyond individual psychic needs, in morality and other cultural formations.[10] Freud's social analysis of the psychic work of jokes (expressing needs and desires) reveals their social aggression and, therefore, political content:

> It is possible to say out loud what these jokes whisper: that the wishes and desires
> of human beings have a right to make themselves heard as much as demanding

Lytton Strachey participated in the British Society for the Study of Sex Psychology, James and Alix Strachey—also trained analysts, members of the London Psychoanalytic Society, and friends and close neighbors of Woolf from the 1920s on—translated Freud's collected papers through the Hogarth Press). Woolf's relationship to Freudian thought moves between conflict and conflicted, from her autobiographical, analytical writings on childhood for the Memoir Club and later interest in Freud's oeuvre to her awful experiences with psychiatrists and horror at the misogynistic aspects of psychoanalysis. Yet, between Woolf's interest in subversive notions of subjectivity, identity, and sexuality and the tremendous influence of Freudian thought on the contemporary moment, it is easy to see how and why critics argue that Woolf engages in an ongoing conversation with Freud, despite—or perhaps in part because of—diary entries and letters like this one: "[We] are publishing all Dr Freud, and I glance at a proof and read how Mr A.B. threw a bottle of red ink on to the sheets of his marriage bed to excuse his impotence to the housemaid, but threw it in the wrong place, which unhinged his wife's mind,—and to this day she pours claret on the dinner table. We could all go on like that for hours; and yet these Germans think it proves something—besides their own gull-like imbecility" (*Letters* vol. 3, 134–5). As if making a case in point for the persistence of critical connections between Woolf and Freud, Nicole Ward Jouve writes, "what is normally done is to see Woolf as refuting Freud rather than—as I see her—both producing alternatives to Freud *but also* open to psychoanalytic readings" (263).

[9]   John Lehman's *Virginia Woolf and Her World* includes an account of the "Dreadnought hoax" and a photo of Virginia and her friends in costume from *The Daily Mirror*, February 16, 1910.

[10]   As we see in *Dialectic of Enlightenment.*

and ruthless morality, and in our times it has been said in forceful and stirring sentences that this morality is only the selfish ordinance of the rich and powerful few who are able to satisfy their wishes without postponement at any time. As long as the art of healing has not gone further in making our life more certain, and as long as social arrangements do not do more to make it more agreeable, the voice in us that rebels against the demands of morality will not be stifled ... only the continued existence of so many unfulfilled demands is able to develop the power to change the social order. (105)

Woolf's turn to fantasy in *Orlando*, and her categorization of the novel as a joke, is indeed political. It is both straightforwardly political (as, for example, a parody of patriarchal property laws) and an attack on "morality," as in this passage in which Orlando changes gender in Constantinople: Our Ladies of Purity, Chastity, and Modesty circle him, now her, and "make as if to cover Orlando with their draperies ... The sisters become distracted and wail in unison, still circling and flinging their veils up and down ... waving their draperies over their heads, as if to shut out something they dare not look upon ..." (96–7).[11] The trumpeters of "Truth" drive away these harpies of Morality: "'We go; we go. I (*Purity says this*) to the hen roost. I (*Chastity says this*) to the still unravished heights of Surrey. I (*Modesty says this*) to any cosy nook where there are ivy and curtains in plenty'" (97).

As the theatrical nature of the previous passage suggests, the H.M.S. *Dreadnought* masquerade is of a piece with *Orlando*. Karen R. Lawrence discusses the novel's use of Constantinople in the context of the "spectacle of Orientalism" (which often included "fancy dress" and cross-dressing) popular in the 1920s:

> The *dual trajectories* of the narrative—centrifugal and liberating, and centripetal and domesticating—create a complex "cultural politics" and poetics. In order to extrapolate the fantasy of sexual boundary crossings, the narrative mines the overdetermined figure of Eastern travel, yet ultimately repatriates the erotic, comic possibilities onto English soil. (256, italics mine)

And yet, Constantinople does not stand in only for the East but also—and I think primarily—as the cultural meeting point of East and West. As such, *Orlando*'s Constantinople subverts traditional Western notions of subjectivity, nation, and empire.[12]

---

[11]   Through the figure of Orlando him/herself, it is also *fantastically* political in the "intergenerational" sense, described here by Freud (excerpted from the previous block quotation): "We must link our lives to that of others in such a way, we must be able to identify with others so closely, that we are able to overcome the curtailment of our own lifetime; and we may not fulfill the demands of our own needs illegitimately, but must leave them unfulfilled, because only the continued existence of so many unfulfilled demands is able to develop the power to change the social order" (105).

[12]   David Roessel's "The Significance of Constantinople in 'Orlando'" places Woolf's connections to Constantinople in the context of its political importance, from Woolf's own experience of the city, Vita's first book of poems (*Constantinople: Eight Poems*), and

The journal record of Woolf's visit to Constantinople in October of 1910 describes the delight of the Turkish landscape viewed from on high (Pera), much as *Orlando* does. Here Woolf engages in a moment of reflection on travel and Western views of Turkey, including the custom of the veil:

> [I]t was not ten years ago that the Turks & Armenians massacred each other in the streets. So perhaps if it were your lot to spend your life here you might think your station one of some happy risk—as a resting place beneath a volcano. Happily a traveler need not trouble himself with the intricate roots of all these strange separate flowers that we look at above ground ... As for those observations upon manners or politics with which all travelers should ballast their impressions, I confess I find myself somewhat out of pocket today. The truth is that travelers deal far too much in such commodities, & my efforts to rid myself of certain preconceptions have taken my attention from the actual facts. Were we not told, for instance, that the female sex was held of such small account in Constantinople—or rather it was so strictly guarded—that a European lady walking unveiled might have her boldness rudely chastised? But the streets are full of single European ladies, who pass unmarked, & that veil which we heard so much of—because it was typical of a different stage of civilization & so on—is a very frail symbol. Many native women walk bare faced; & the veil when worn is worn casually, & cast aside if the wearer happens to be curious. (Morris 231–2)

Here Woolf acknowledges that culture, politics, and violence are deep-rooted phenomena, and that travelers tend to stay "above ground," on the surface of places. For this reason, their attempts at cultural commentary are mere "ballast." Even a self-conscious traveler like Woolf, who wishes to sail without ballast, may miss what passes before her eyes in the process of unloading such excess baggage. The veil is, in this instance, a multivalent symbol, of a "frail" patriarchy and patriarchal morality (also wielded by Our Ladies of Purity, Chastity, and Modesty) and of a traveler's shrouded or changing vision.

As the most astute contemporary review of the novel, Rebecca West's "High Fountain of Genius," argues the novel combats the determinism of realism with the truth of comedy and fantasy, combining "the frankest contempt for realism with the profoundest reality" (*New York Herald Tribune*, October 21, 1928).[13] *Orlando's* fantasy expresses the "profoundest reality" not only in opposition to the realism of writers such as Arnold Bennett, as West argues, but also the unreason of liberal humanism and modernity. Four centuries of garden and desert whirl fantastically around the title character, yet this fantastic movement foregrounds the place of place in the constitution of the subject and processes of subjectivity.

---

associations of Turkey with Sapphism to Leonard Woolf's 1917 monograph, "The Future of Constantinople" and the Chanak crisis of 1922. As noted in the last chapter, Shelley named her young Turkish woman "Safie."

[13] Makiko Minow-Pinkey briefly refers to Freud's theory of jokes as "the truth of the unconscious" in her argument for *Orlando*'s value as a serious work of art (117).

The Place of Place

> [S]o much of Virginia Woolf's writing is travel writing in her own kind. Few writers have ever been more powerfully inspired by the sense of place ... Virginia Woolf was not a spectacular traveler, nor a natural wanderer ... except for a fleeting visit to Asiatic Turkey in 1910 she never went out of Europe.
>   —Jan Morris, *Travels with Virginia Woolf*

> At six I was on deck, & suddenly we found ourselves confronted with the whole of Constantinople; there was St Sophia, like a treble globe of bubbles frozen solid, floating out to meet us. For it is fashioned in the shape of some fine substance, thin as glass, blown plump curves; save that it is also substantial as a pyramid. Perhaps that may be its beauty. But then beautiful & evanescent & enduring, to pluck adjectives like blackberries—as it is, it is but the fruit of a great garden of flowers.
>   —Virginia Woolf, *Journal*, Greece, October 1906

Unlike Candide and the monster, Orlando does not wander the world. He leaves his idyllic surroundings for Turkey because his ancestral house becomes, as the biographer-narrator writes, "uninhabitable" (82). Orlando travels not only to distance himself from what has been the symbol of his identity and Englishness, but also to flee the specter of marriage and other conventions.

Yet, as the English Ambassador to Constantinople, Orlando finds himself again lost among duty and tradition. Preferring the company of dogs and mountain tops, he is often heard by local shepherds reciting his English poem, "The Oak Tree," the manuscript of which he has been carrying and composing for well over a hundred years. When we first meet Orlando he is in the process of writing another poem, "Aethelbert: A Tragedy in Five Acts," and having great difficulty with the fact that "Green in nature is one thing; green in literature another. Nature and letters seem to have a natural antipathy; bring them together and they tear each other to pieces" (13). In Constantinople, the shepherds mistake his sing-song recitations for prayer to his god, a mistake which suggests more than simple irony.

Orlando seems to "go native," learning the local language, wearing native dress, and loving the landscape. But, to the locals, it is this very love of landscape that marks him as foreign (and, to the narrative-biographer, as English: the "English Disease"):

> That he, who was of English root and fibre, should yet exult to the depths of his heart in this wild panorama, and gaze and gaze at those passes and far heights planning journeys there alone on foot where only the goat and shepherd have gone before; should feel a passion of affection for the bright, unseasonable flowers, love the unkempt, pariah dogs beyond even his elk-hounds at home, and snuff the acrid, sharp smell of the streets eagerly into his nostrils, surprised him. (85)

The metaphor here—"root and fibre"—is that of the English Oak transplanted. Orlando is not only writing "The Oak Tree," he is the oak tree. Unlike the great house, the other symbol of his identity, the oak tree embodies living processes, doubled by Orlando's continuously composed poem as a narrative of his travel and development (in this sense the "The Oak Tree," like the monster's narrative, might be thought of as another embedded Bildungsroman, though we see only a few lines of the actual text).

In Constantinople, Orlando yearns to be away from his official "duty"—duty is, after all, nothing more that the voice of acculturation demanding "remember who you are," meaning who one is expected to be—and closer to the landscape, the desert earth. "Often [he] ... had looked at those mountains [outside Broussa] from [his] ... balcony at the Embassy; often had longed to be there" (99). This yearning for the mountains rather mysteriously changes Orlando the man into Orlando the woman. One evening he receives his Dukedom and the Order of the Bath and wakes, a week later,[14] a woman—sleeping deeply through an uprising against the Sultan, avoiding the masculine duty of battle altogether. She leaves the city, as if pre-arranged, on a donkey led by an old gypsy. In this journey to become oneself, Orlando sheds one gender for another and one idea of self (as tradition or duty) for another (as experience).

While in the gypsy camp, the community suspects Orlando of worshipping nature, a grave offense in their eyes:

> The English disease, a love of nature, was inborn in her, and here, where Nature was so much larger and more powerful than in England, she fell into its hands as she had never done before ... She climbed the mountains; roamed the valleys; sat on the banks of the streams. She likened the hills to ramparts, to the breasts of doves ... Trees were withered hags and sheep were grey boulders. *Everything, in fact, was something else* ... and when, from the mountain-top, she beheld far off, across the sea of Marmara, the plains of Greece, and made out ... the Acropolis with a white streak or two which must, she thought, be the Parthenon, her soul expanded with her eyeballs, and she prayed that she might share the majesty of the hills ... as all such believers do. (101, italics mine)

Orlando's perception of the landscape—as a large, powerful manifestation of nature—isn't simply an English exoticization of Broussa; it is a particular way of seeing.[15] Her way of seeing, Orlando's response to her environment, is metaphor itself ("Everything ... was something else"). Perception leads her to an idea of beauty and from there to an idea of truth:

---

[14] In Jamaica Kincaid's Bildungsroman *Annie John*, Annie also suffers a mysterious, transformative illness, characterized by excessive sleep, though it seems less like a fairy-tale trope in Kincaid's text.

[15] Not unlike the process Ruskin enacts in *The Stones of Venice*; on Virginia's first trip to Italy, her older sister Vanessa spent a good part of the trip finding fault with Ruskin's book (Bell 88).

> She began to think, was Nature beautiful or cruel; and then she asked herself
> what this beauty was; whether it was in things themselves, or only in herself; so
> she went on to the nature of reality, which led her to truth, which in turn led to
> Love, Friendship, and Poetry (as in the old days on the high mound at home);
> which meditations ... made her long ... for pen and ink. (102)

For the Romantics, mountains embodied the idea of nature. From the "high mound at home" to the Turkish mountains, Orlando seems engaged in a Keatsian pursuit of Beauty and Truth (with the Turkish landscape, and the view of Greece, another Urn). Longing for pen and ink, she makes ink from berries and wine, and writes in shorthand in the margins of her manuscript a description of "the scenery in a long, blank verse poem, and to carry on a dialogue with herself about this Beauty and Truth concisely enough. This kept her extremely happy for hours on end" (102–3). This way of seeing is not without drawbacks; the gypsies agree that "here is someone who does not do the thing for the sake of doing; nor looks for looking's sake; here is someone who believes neither in sheep-skin nor basket; but sees (here they look apprehensively about the tent) something else" (103).

And yet, metaphor might also signify, or in the right circumstances become, a way of seeing that reaches through what Woolf called "the cotton wool of daily life" toward the "revelation of some order," that is, "some real thing behind appearances," in her autobiographical essay, "A Sketch of the Past" (72). In other words, looking at a sheep-skin and seeing "something else," such as the grass and air and light and water and interconnected social and physical processes that constitute it, isn't necessarily not seeing the skin or using it as canvas for human desire. It *may* be a way of seeing more. From this perspective, a certain use of metaphor may be a way to describe what Stacy Alaimo calls the transcorporeality of all bodies, as ecosystems in their own right and as bodies connected materially, cellularly, to other larger ecosystems. "*Imagining* human corporeality as trans-corporeality, in which the human is always intermeshed with the more-than-human world, underlines the extent to which the substance of the human is ultimately inseparable from 'the environment'" (2, italics mine). Like Orlando's fluid metamorphosis of gender, metaphor in this sense may be a metamorphosis of perspective.

Near the end of the book, in the twentieth century, Orlando returns to her ancestral home as a woman and, unlike her real-life counterpart Vita, is allowed to keep her house (despite being legally dead and a woman). Once again Orlando returns to the oak tree on the hill that overlooks her grand estate, where we found him at the beginning of the novel, pondering the question of the relationship between nature and letters. Here, at the end of her journey, Orlando stands ready to bury her nearly four-hundred-year-old manuscript of "The Oak Tree," now a published work, at the base of its referent. However, as the ground is too hard, she concludes that "[n]o luck ever attends ... symbolic celebrations" (225), questioning the task at hand and the relationship between nature and letters:

So she let her book lie unburied and disheveled on the ground, and watched the vast view, varied like an ocean floor ... In the far distance Snowdon's crags broke white among the clouds ... 'And there,' she thought, letting her eyes, which had been looking at these far distances, drop once more to the land beneath her, 'was my land once; that castle between the downs was mine; and all that moor running to the sea was mine.' Here the landscape ... shook itself, heaped itself, and let all this encumbrance of houses, castles, and woods slide off its tent-shaped sides. The bare mountains of Turkey were before her. It was blazing noon. She looked straight at the baked hill-side ... At this moment some church clock chimed in the valley. The tent-like landscape collapsed and fell. The present showered down upon her head once more ... (225–6)

While Orlando's vision is at first appropriative (an unnaturally expansive, imperial vision, as it was in Broussa when she could see the Acropolis), taking in the crags of Snowdon far beyond the normal range of sight, it becomes a vision of land itself as an agent, a site of experience. It is a vision of human vision itself as an embedded, interconnected process, which has onto-epistemological, ethical, and political implications.[16] As Alaimo argues, "Understanding the material world as agential and considering that things, as such, do not precede their intra-actions are, I think, crucial ... the existence of anything—any creature, ecosystem, climatological pattern, ocean current—cannot be taken for granted as simply existing out there" (21).

Travel foregrounds place itself (both inside and outside the body) as the multiplicity of experience. Orlando's travels through four centuries of time and space form a movement away from a perception of the world and self as static, self-contained units, to a perception of both as interconnected forms of subjectivity, as sites of experience. In the sixteenth century, Orlando seems set to make his mark on the world and others, as his fathers before him, "slicing at the head of a Moor which swung from the rafters ... [a head that] Orlando's father, or perhaps his grandfather had struck it from the shoulders of a vast Pagan" (11).[17] By the eighteenth century, she is an Aeolian Harp played by the winds of destiny: "[Orlando] became conscious, as she stood at the window, of an extraordinary tingling and vibration all over her, as if she were made of a thousand wires upon which some breeze or errant fingers were playing scales" (165).[18] Finally, in the twentieth century Orlando becomes neither a subject capturing objects (in life or

---

[16] Indeed, "ethical considerations and practices must emerge from [this] ... more uncomfortable and perplexing place where the 'human' is always already part of an active, often unpredictable, material world" (Alaimo 16–17).

[17] This scene, recalled at the end of novel, explicitly contrasts the Crusades with Orlando's travels, killing "natives" with "going native."

[18] Later this Harp becomes *almost* an Angel in the House: "Thus did the spirit [of the century] work upon her ... those twanglings and tinglings which had been so captious and so interrogative modulated into the sweetest melodies, till it seemed as if angels were plucking harp strings with white fingers and her whole being was pervaded by a seraphic harmony" (168).

language) nor the object of worldly or otherworldly spirits. In the present moment of 1928 she calls out to herself, "hesitatingly, as if the person she wanted might not be there, 'Orlando?'" and is answered with the revelation of not one or two or twenty other selves but, possibly, thousands:

> these selves of which we are built up, one on top of another, as plates are piled on a waiter's hand, have attachments elsewhere, sympathies, little constitutions and rights of their own ... [Orlando] had a great variety of selves to call upon, far more even than we have been able to find room for, since a biography is considered complete if it merely accounts for six or seven selves, whereas a person may well have as many thousand. (212–13)

As the biography of a traveler sensitive to place, the novel here admits its own impossibility—how can a biography encompass several thousand selves? Orlando's search for "the Captain self, the Key self, which amalgamates and controls them all" (214) is "the wild goose. It flies past the window and out to sea. ... I've seen it, here—there—there—England, Persia, Italy. Always it flies fast out to sea and always I fling after it words like nets (here she flung her hand out) which shrivel" (216). When she finally ceases to call for it, this goose, "what is called, rightly or wrongly, a single self, a real self' comes to her (216) only to, as we shall see, reappear in flight at the end of the book. It suggests the impossibility, the unreality of the self-contained subject of liberal humanism and the view of human and nonhuman nature it relies on.

Here epiphany seems like the trumpets of "Truth" in Constantinople, a sword to cut through patriarchal morality, convention, and culture. Unlike the sword Orlando wields when we first meet him, swinging at a human head hanging from the rafters of his house, this axe breaks the frozen sea within culture.[19] This epiphany is close to intuition, to the personal insight into the world, which is also political. It is the revelation of multiple selves that connects to Orlando's vision of vision itself, to the interconnected multiplicities of the self and world.

Woolf records her first encounter with this epiphany, in her childhood garden at St. Ives: "'That is the whole,' I said. I was looking at a plant with a spread of leaves; and it seemed suddenly plain that the flower itself was a part of the earth; that a ring enclosed what was the flower; and that was the real flower; part earth; part flower" ("A Sketch" 71). In this way, St. Sophia and Constantinople itself (and Orlando, the Turkish mountains, the "high mound" at home) are, as Woolf's journal entry of October 1906 states, "but the fruit of a great garden of flowers"[20]—part of an interconnected whole. And yet, the *voice of tradition*, the liberal humanist biographer, commands that the "first duty" of a biographer, historian, or traveler is to avoid fruit and flower, to avoid intuitive, transformative experience of the subject and the world: "the first duty of a biographer ... is to

---

[19]   Kafka, also quoted in the Introduction: "Literature is the axe to break the frozen sea within us."

[20]   Quoted as an epigraph to this section.

plod, without looking to right or left, in the indelible footprints of truth; unenticed by flowers; regardless of shade; on and on methodically till we fall plump into the grave and write *finis* on the tombstone above our heads" (*Orlando* 47).

Biography, Biology, and the Bildungsroman

> [T]he renewed interest in memoirs [in the 1920s] is biological: and perhaps the portraitist of to-day, who is first of all a psychologist, is much nearer to the biologist than the historian.
>
> —Emil Ludwig, *Genius and Character*

> [B]iography can seem to have less in common with puzzle-built pictures of heroes than a visit to Frankenstein's laboratory.
>
> —Miranda Seymour, "Shaping the Truth"

> Human corporeality, especially female corporeality, has been so strongly associated with nature in Western thought that it is not surprising that feminism has been haunted not only by the specter of nature as the repository of essentialism, but by, as Linda Birke puts it, "the ghost of biology."
>
> —Stacy Alaimo, *Bodily Natures*

In "The Proper Study?" Richard Holmes, intrepid biographer of Percy Shelley (among others), twice quotes Dr. Johnson's essay "On Biography" from *The Rambler*[21]: "[N]o species of writing seems more worthy of cultivation than biography, since none can be more delightful or more useful, none can more certainly enchain the heart by irresistible interest, or more widely diffuse instruction to every diversity of condition" (Holmes 11). Holmes reads Johnson's promotion of biography as part of the generally subversive project of acknowledging a common human nature, common to criminals and kings (12–13).[22] Holmes goes on to argue that biography itself is a unique cultural index; indeed, scholars of biography "would discover how reputations developed, how fashions changed, how social and moral attitudes moved, how standards of judgment altered, as each generation, one after another, continuously reconsidered and idealized or condemned its forebears in the writing and rewriting of biography" (17–18).

Orlando takes up Johnson's view that biography reveals our "common nature," at once contesting, as we have seen, the humanist idea of a bounded, discreet (universal) self *and* claiming that not only criminals and kings, but all men and women share qualities and desires: both "genders" are androgynous. Like Holmes,

---

[21] No. 60, 1750.

[22] In this vein, Holmes remarks, "[I]n 1817 Mary Shelley chose to educate Frankenstein's Monster in the complex ways of human civilization by making him read biography ('a volume of Plutarch's Lives') as well as Goethe's fashionable novel *The Sorrows of Young Werther* ..." (13–14). While fiction depresses the monster, biography edifies as a window to history and philosophy.

Woolf also believes in the force and value of biography, not simply as a didactic tool but as a window into culture. In "The New Biography" Woolf writes, in the twentieth century

> The point of view [of biography] had completely altered ... [T]he author's relation to his subject is different. He is no longer the serious and sympathetic companion, toiling even slavishly in the footsteps of his hero. Whether friend or enemy, admiring or critical, he is an equal. In any case, he preserves his freedom and his right to independent judgment. Moreover, he does not think himself constrained to follow every step of the way. Raised upon a little eminence which his independence has made for him, he sees his subject spread about him. He chooses; he synthesizes; in short, he has ceased to be the chronicler; he has become an artist. (475)

"Raised upon a little eminence" this new biographer sounds a lot like the Romantic artist, like Orlando himself. Perhaps the most striking difference between the Victorian and the "new" biography[23] is the attempt to capture what Woolf calls the "rainbow" of intangibles (such as personality) as well as the "granite" of fact; "the life which is increasingly real to me is the fictitious life; it dwells in personality rather than in the act. Each of us is more Hamlet, Prince of Denmark, than he is John Smith, of the Corn Exchange" ("The New Biography" 478). With *Orlando*, it seems that Woolf decided that the Bildungsroman—as the fictional analogue of biography[24]—is uniquely suited to this attempt. Indeed, Katherine Miles refers to *Orlando* as Woolf's "fictional praxis" of her theory of biography (213).

If, as Woolf insists, *Orlando* is a joke, what is this "serious joke" about? Like the Bildungsroman, biography occupied a pedagogical role in Victorian culture. The exemplary lives of exemplary men were recorded and read as lessons in virtue and success. William Butler Yeats's assertion that all history (or, as some have it, all knowledge) is biography certainly expresses this view. And yet, there is a sense in which Yeats's phrase signifies more than Victorian convention. All of history is biography in as much as history is the story—not of great men, as

---

[23]   Kay Ferres's "Gender, Biography, and the Public Sphere" notes that concepts of "biography and its public uses figure in all three [*Three Guineas, Roger Fry*, 'A Sketch of the Past']; and in all of them Woolf considers the workings of influence: how individuals reproduce and transform public culture" (313). Hermione Lee also views *A Room of One's Own* and *Three Guineas* to be essays on "life-writing" (15).

[24]   As Elinor S. Shafer notes, "the first full-scale book on biography, *An Essay on the Study and Composition of Biography*," was published in 1813, "almost at the same time as the word *Bildungsroman* was coined in Germany by Karl von Morgenstern" (116). She goes on to argue that, "the *Bildungsroman* stands as an 'ideal' form which could resolve the clash between unvarnished fact and edification. The power to shape and transform the facts through an imaginative portrayal, which Coleridge and Froude reserved for religious figures and Shakespeare, was in the course of the nineteenth century accorded, not without struggle, to the new dominant form, the novel, in particular the *Bildungsroman*, and its near relation, the biography" (133–4).

Carlyle wrote—but of the minds of those writing history. Just as the Victorian biographer absents himself from biography, so the traditional historian adopted the voice of an objective recorder of fact. However, Woolf makes biography itself one of the subjects of her novel, literally and figuratively; the other main character of *Orlando: A Biography* is the biographer him/herself. Here, the double-facedness of this joke, as in all jokes, lies in its relation to audience. Leslie Stephen, Woolf's brilliant and domineering father, was one of the principle architects and a long-time editor of the *Dictionary of National Biography*,[25] and a friend and colleague of Carlyle, two figures with whom Woolf is in conversation all her life. *To the Lighthouse*, published just one year before *Orlando*, is a feminist, Modernist critique of the Victorian man of letters. As many have noted, Mr. Ramsey strongly resembles Woolf's father: "He was incapable of untruth; never tampered with fact; never altered a disagreeable word to suit the pleasure or convenience of any mortal being, least of all his own children, who, sprung from his loins, should be aware from childhood that life is difficult; facts uncompromising" (10–11).[26] Leon Edel remarks, "Stephen based his editorial principles on a belief in the primacy of facts. Ideas and the discussion of ideas had no place, he argued, in encyclopedic biographical accounts" (*Writing Lives* 71). Heavily autobiographical, *To the Lighthouse* reworks childhood memories at St. Ives in light of Woolf's adult knowledge and desires[27]—the desire to make one's destiny in the face of human limitations and moral claptrap, specifically to make art despite the dictates of patriarchy. One could even think of *Orlando* as the novel that would be written by Lily Briscoe (were she a writer and not a painter, like Woolf's sister) or by Nancy

---

[25]   Stephen edited the *Dictionary of National Biography* for sixteen years from 1885.

[26]   As I will argue elsewhere, he is also, at times, an embodiment of the persona of Mathew Arnold's "Dover Beach." The poem is alluded to repeatedly but, unlike "The Oak Tree" in *Orlando*, never quoted in the novel. Though a remarkable number of poems are quoted in *To The Lighthouse* (Charles Elton's "Luriana Lurilee," Shakespeare's "Sonnet 98," William Browne's "Siren Song," William Cowper's "The Castaway," and Tennyson's "The Charge of the Light Brigade") "Dover Beach" seems central to the text, much the way that "The Oak Tree" is, as Dudley Marchi argues, the absent centerpiece of *Orlando* (more on this in a moment).

[27]   Interestingly, St. Sophia and Constantinople appear in *To the Lighthouse* as well, as metaphor: "What was it she wanted? Nancy asked herself. There was something, of course, that people wanted; for when Minta took her hand and held it, Nancy, reluctantly, saw the whole world spread out beneath her, as if it were Constantinople seen through a mist, and then, however heavy-eyed one might be, one must needs ask, 'Is that Santa Sofia?' 'Is that the Golden Horn?' So Nancy asked, when Minta took her hand. 'What is it that she wants? Is it that?' And what was that? Here and there emerged from the mist (as Nancy looked down upon life spread beneath her) a pinnacle, a dome; prominent things, without names" (112–13). And Constantinople appears again, this time as an image in Cam's mind: "And the drops falling from this sudden and unthinking fountain of joy fell here and there on the dark, the slumbrous shapes in her mind; shapes of a world not realised but turning in their darkness, catching here and there, a spark of light; Greece, Rome, Constantinople" (281).

Ramsey (who, arguably, is a figure for Woolf herself) when she's older. With *Orlando*, then, Woolf's audience is both historical (patriarchy, the "great men" of history, and the Victorian cult of virtue) and contemporary (her literary world and the reading public). The "joke" of the novel is about, indeed is at the expense of, this personal and public "his-story" (her father, history, literary tradition, and patriarchal pedagogy and culture generally).

Orlando is a "her-story" for these reasons and, of course, because it is a magical or mock-biography of Sackville-West. Again, though the novel challenges Freud's view that "anatomy is destiny," as fantastic wish-fulfillment it meets his injunction that we "link our lives to that of others in such a way ... that we are able to overcome the curtailment of our own lifetime" leaving the "demands of our own needs ... unfulfilled, because only the continued existence of so many unfulfilled demands is able to develop the power to change the social order" (*Jokes* 105).[28] The novel dramatizes the "continued existence of unfulfilled demands" with respect to gender roles, laws of property, and the propagation of war. Explicitly, Woolf attacks the idea that anatomy is deterministic not only through Orlando's sex change but through the satirical depiction of constructions of gender across the centuries (in terms of dress, deportment, legal rights, etc.).

Quentin Bell's biography of Woolf[29] traces *Orlando* as just this sort of magical or metaphorical biography of Sackville-West. Bell notes the many connections between the novel and real incidents in Virginia's life in the 1920s: "Vita at Knole, showing her over the building—4 acres of it—stalking through it in a Turkish dress surrounded by dogs and children; a cart bringing in wood as carts had done for centuries to feed the great fires of the house; Vita hunting through her writing desk to find a letter from Dryden" (132). And, of course, like Orlando, Vita was a poet, "courted and caressed by the literary world; the homage of Sir Edmund Gosse, and indeed of Virginia herself" (132), and her poem "The Land" the model for "The Oak Tree."[30] Sandra M. Gilbert furthers this list:

> her early impassioned affair with Violet Trefusis (who here becomes the Russian
> Sasha because Vita called her Lushka); her Spanish grandmother (here, as in
> real life, 'Rosina Pepita'); her courtship by the foolish aristocrat Lord Lascelles
> (here the Duke/Duchess of Scand-op-Boom); her travels in the East (here the
> Turkish episode); her transvestism (here the eighteenth-century escapades);
> her winning of the Hawthornden Prize for 'The Land' (here the winning of
> the 'Burdett Coutts Prize' for 'The Oak Tree'); her legal fight for her ancestral

---

[28] Freud himself was obsessed with biography, writing biographies of Leonardo da Vinci and Woodrow Wilson (as well as many detailed case-studies). Of course, Freud's oeuvre overcomes "curtailment" to take a wider view. As Malcolm Bowie writes, "Freud himself has a part to play in his *Bildungsroman* of the human species" (see "Freud and the Art of Biography").

[29] *Virginia Woolf: A Biography*.

[30] As critics Frank Baldanza and Charles G. Hoffmann point out, the few verses of "The Oak Tree" that appear in the novel are, in fact, from Vita's "The Land."

property (here, as in real life, her Great Law Suit); her marriage to the supportive bisexual Harold Nicholson (here called Marmaduke Bonthrop Shelmerdine); and so forth. (xxviii–xxix)[31]

In addition, Shelmerdine's name builds on Vita's nickname for Harold, "Mar," and, like Vita, Orlando gives birth to a boy. While the connections here are many, the novel's concerns with biography fall on a much grander scale than even that of the colorful life and ancestry of Sackville-West.

On March 18, 1928 Woolf writes, "I have written this book quicker than any: & it is all a joke; & yet gay & quick reading I think: a writer's holiday" (*Diary* vol. 3, 177). This writer's "holiday" is also a journey through four centuries of English literary history—at the heart of which is Orlando's enduring composition, "The Oak Tree." In fact, Dudley M. Marchi reads the novel itself as the story of the composition of the poem, which ends when it is complete; although we never get to read "The Oak Tree," "it is the missing centerpiece of *Orlando*" (21). Though this tree symbolizes the grand lineage of England (literary and otherwise), there is more to this omission than the subversion of the Victorian convention of life and letters. Just as Candide's garden signifies an actual, material garden, "The Oak Tree," as the absent referent of *Orlando*, signifies explicitly the impossibility of "capturing" nature in letters. Negatively, however, we may read this pointed absence of text as the presence of the more-than-human world—the long "shadow" of biology cast by every biography. The narrator hints that "nature" is an active participant in the composition of "The Oak Tree" and, presumably, *Orlando* itself: "The shade of green Orlando now saw spoilt his rhyme and split his meter. Moreover, nature has tricks of her own. Once look out of a window at bees among flowers, at a yawning dog, at the setting sun ... and one drops the pen ..." (13).

What of the long shadow cast by the "actual" oak tree represented in *Orlando*? In *Frankenstein*, the shadow of biology is also figured, however briefly, in an oak tree, the tree that is split violently by lightening, fomenting Victor's scientific epiphany. Here, the "actual" oak tree near Orlando's ancestral home *is* the epiphany. Persevering with the character through the ages, the tree *is* Orlando: ancestral home, noble lineage, cyclic nature. The fraction of Orlando's life recorded in the novel is an unnaturally long period for a human, but a relatively reasonable one for an oak. "The Oak Tree" is both an absent poem about Orlando's relationship to the more-than-human world *and* the presence of this world in and around the text. Here, biography is also biology.

Biology constitutes the self *and* the other—constitutes but does not determine; culture is a part of nature, and nature and culture are co-creating processes. Woolf addresses the feminist fear of nature and biology as, again as Alaimo has it, "the repository of essentialism" (5) with an ecological vision of human consciousness. One year before *Orlando*, three key moments in *To the Lighthouse*, each from the perspective of a central female character, suggest the deep material (including the

---

[31]   See Gilbert's introduction to the Penguin edition of the novel.

seemingly ethereal) interconnection of human and nonhuman nature. First, Mrs. Ramsey: "It was odd, she thought, how if one was alone, one leant to inanimate things; trees, streams, flowers; felt they expressed one; felt they became one; felt they knew one, in a sense were one; felt an irrational tenderness thus (she looked at that long steady light) as for oneself" (97–8). This alludes to Percy Bysshe Shelley's "On Love,"[32] but with a difference. Mrs. Ramsey and the trees, streams, and flowers are, "in a sense ... one." Here, "oneself" is another material entity; *if* trees are "inanimate," then Mrs. Ramsey's language suggests that we too have our lives as objects, part of the atoms and fibers of the world. The tenderness we feel for trees and streams is, then, just as irrational or rational as our feeling for ourselves. Second, Nancy Ramsey: "Brooding, she changed the pool into the sea, and made the minnows into sharks and whales, and cast vast clouds over this tiny world by holding her hand against the sun, and so brought darkness and desolation, like God himself, to millions of ignorant and innocent creatures, and then took her hand away suddenly and let the sun stream down" (114–15). Here, Nancy is hypnotized, awe-struck by the interconnectedness of scale, of the vastness and tininess of life (again, Woolf's pattern behind "the cotton wool"), and "the intensity of feelings which reduced her own body, her own life, and the lives of all the people in the world, for ever, to nothingness" (115). Finally, the artist Lily Briscoe: "She raised a little mountain for the ants to climb over. She reduced them to a frenzy of indecision by this interference in their cosmogony. Some ran this way, others that" (294). This scene leads Lily, in the next paragraph, to wish for a "secret sense" with which to understand other creatures, including other human beings—a godlike or novelist's consciousness. "One wanted most some secret sense, fine as air, with which to steal through keyholes and surround her [Mrs. Ramsey] where she sat knitting, talking, sitting silent in the window alone; which took to itself and treasured up like the air which held the smoke of the steamer, her thoughts, her imaginations, her desires. What did the hedge mean to her, what did the garden mean to her, what did it mean to her when a wave broke?" (294).

These last two examples might, in another context, seem to emphasize human power but here they do not—Nancy understands that she and her species are like the tidal pool animals, subject to forces beyond her control (be they internal or external). Lily's "interference" in ant cosmogony also casts her play as god-like, but it too is followed by the realization of human limitations, expressed as the desire for a "secret sense, fine as air"—a sense that well describes the narrative presence of the "Times Passes" section of the novel. These scenes, far from affirming the

---

[32] "This is Love. This is the bond and the sanction which connects not only man with man, but with everything which exists. We are born into the world, and there is something within us which, from the instant that we live, more and more thirsts after its likeness. ... Hence in solitude, or in that deserted state when we are surrounded by human beings, and yet they sympathize not with us, we love the flowers, the grass, and the waters, and the sky. In the motion of the very leaves of spring, in the blue air, there is then found a secret correspondence with our heart." From *Essays, Letters from Abroad, Translations and Fragments,* 2 vol. Ed. Mary Shelley. London: Edward Moxon, 1840.

Victorian view of a hierarchical order in nature, challenge it by demonstrating how our consciousness of ourselves and others is shaped by, and shapes, a complex material world, human and nonhuman, animate and "inanimate."

Let's return again to one of the key moments in *Orlando* suffused with the "English disease," a love of nature: in Broussa, Orlando "climbed the mountains; roamed the valleys; sat on the banks of the streams. She likened the hills to ramparts, to the breasts of doves ... Trees were withered hags and sheep were grey boulders. Everything, in fact, was something else" (101). In light of *To the Lighthouse*, does this passage still seem, as it does at first glance, *only* an appropriative gesture, a metaphorical taxonomy? Or might it be read, as I suggested earlier, as a movement toward seeing the material connectedness of all things? Consider it alongside this passage from another of Woolf's unconventional biographies/Bildungsromane, *Flush*, published in 1933, in which the title character (Elizabeth Barrett Browning's cocker spaniel) first enters Elizabeth's bedroom:

> In the middle of the room swam up to the surface what seemed to be a table with a ring around it; and then the vague amorphous shapes of armchair and table emerged. But everything was disguised. On top of the wardrobe stood three white busts; the chest of drawers was surmounted by a bookcase; the bookcase was pasted over with crimson merino; the washing-table had a coronal of shelves upon it; in top of the shelves that were on top of the washing-table stood two more busts. *Nothing in the room was itself; everything was something else.* Even the window-blind was not a simple muslin blind; it was a painted fabric with a design of castles and gateways and groves of trees, and there were several peasants taking a walk. Looking-glasses further distorted these already distorted objects so that there seemed to be ten busts of ten poets instead of five; four tables instead of two. (20, italics mine)

It is not only the dim light of the sick room that causes Flush's confusion; it is the atmosphere of mimesis: art and mirrors. Flush's impressions of human culture are not unlike Orlando's impressions of nonhuman nature in Broussa. Is Flush expressing his culture, his view of the materiality of our culture, or human culture second-hand? And where would these lines begin?[33]

Woolf's account of the connections and differences between Flush and Barrett Browning reads here as a case of cultural difference:

> And yet sometimes the tie would almost break; there were vast gaps in their understanding. Sometimes they would lie and stare at each other in blank bewilderment. Why, Miss Barrett wondered, did Flush tremble suddenly, and whimper and start and listen? She could hear nothing; there was nobody in the room with them. She could not guess that Folly, her sister's little King Charles, had passed the door; or that Catiline, the Cuba bloodhound, had been given

---

[33] What we find alarming—confusion, distortion, multiplication—is also frightening to Flush, or rather, to Woolf's representation of the spaniel who lived with Barrett Browning (in part a representation of Woolf's own dog, Pinka, a gift from Sackville-West).

a mutton-bone by a footman in the basement. But Flush knew; he heard; he was ravaged by the alternate rages of lust and greed. Then with all her poet's imagination, Miss Barrett could not divine what Wilson's wet umbrella meant to Flush; what memories it recalled ... Flush was equally at a loss to account for Miss Barrett's emotions. There she would lie hour after hour passing her hand over a white page with a black stick; and her eyes would suddenly fill with tears; but why? ... But there was no sound in the room, no smell to make Miss Barrett cry. Then again Miss Barrett, agitating her stick, burst out laughing ... He could smell nothing; he could hear nothing. There was nobody in the room with them. The fact was that they could not communicate with words, and it was a fact that led undoubtedly to much misunderstanding. Yet did it not also lead to a peculiar intimacy? ... Can words say anything? (36–7)

This is a depiction of cultural difference, but the difference of cultures that overlap. For example, both recognize the presence of signs he or she cannot detect and the reciprocity of creating a joint understanding, however unclear and tentative. Here, in their connection to each other, both are alive to the presence of thought's varied material processes, of each other as real subjects, as each observes: "There was nobody in the room with them."

The Wild Goose

Woolf relates her epiphany that the flower is part of the earth, quoted earlier from "A Sketch of the Past," in the context of a discussion of the nature of thought and being. An awareness of the interconnectedness of all things comes to Woolf in three scenes, all of which take place in the garden at St. Ives: a violent fight with her brother Thoby, a meditation on a flower, and a momentary connection between the suicide of a family friend (discussed in her presence by her parents) and an apple tree. The first and third epiphanies filled Woolf with horror—the pain of the world almost rendered her passive and helpless: "It was as if I became aware of something terrible; and of my own powerlessness" (71). The second, however, brought a keen satisfaction, a sense of her active part in the processes of life, intuiting and naming "the whole" (71). All three instances express a deep awareness of others (in the case of other people, of their pain and joy) as part of the self. These moments seem to emanate as much from without as from within:

[T]hough I still have the peculiarity that I receive these sudden shocks, they are now always welcome; after the first surprise, I always feel instantly that they are particularly valuable ... I feel that I have had a blow; but it is not, as I thought as a child, simply a blow from an enemy hidden behind the cotton wool of daily life; it is or will become a revelation of some order; it is a token of some real thing behind appearances; and I make it real [for myself] by putting it into words. ... From this I reach what I might call a philosophy; at any rate *it is a constant idea of mine; that behind the cotton wool is hidden a pattern; that we—I mean all human beings—are connected with this; that the whole world is a work of art; that we are parts of the work of art ... This intuition of mine—it*

is so instinctive that is seems given to me, not made by me—has certainly given
its scale to my life ever since I saw the flower in the bed by the front door at St.
Ives. (72, italics mine)

This instinct or philosophy seems, in fact, both given and made, inherent in
the world as "nature" and "culture." The truth of human culture is its origin in
and continuity with the larger natural world; Woolf's awareness of the deep
interconnectedness or transmateriality of life is itself an example of the way in
which the nonhuman world impacts, shapes, human culture.

In *Orlando*, epiphany arrives in the same way, from without as from within.
It is a moment in which the character (and narrator) senses that the nonhuman
world may intervene in human history. At the end of novel, after the "tent-like
landscape" falls, the present showers upon her and Orlando's inner and outer
vision interpenetrate:

> It was not necessary to faint now in order to look deep into the darkness where
> things shape themselves and to see in the pool of the mind now Shakespeare,
> now a girl in Russian trousers, now a toy boat on the Serpentine, and then the
> Atlantic itself, where it storms in great waves past Cape Horn. There was her
> husband's brig, rising to the top of the wave! Up it went, and up and up. The
> white arch of a thousand deaths rose before it. Oh rash, oh ridiculous man,
> always sailing, so uselessly, round Cape Horn in the teeth of a gale! ... And
> then the wind sank and the waters grew calm; and she saw the waves rippling
> peacefully in the moonlight.
>
> 'Marmaduke Bonthrop Shelmerdine!' She cried, standing by the oak tree.
>
> The beautiful, glittering name fell out of the sky like a steel-blue feather. She
> watched it fall, turning and twisting like a slow-falling arrow that cleaves the
> deep air beautifully. He was coming, as he always came, in moments of dead
> calm ... the moon was on the waters, and nothing moved between sky and sea.
> Then he came.
>
> ... The aeroplane rushed out of the clouds and stood over her head ... And as
> Shelmerdine, now grown a fine sea-captain, hale, and fresh-coloured, and alert,
> leapt to the ground, there sprang up over his head a single wild bird.
>
> 'It is the goose,' cried Orlando. 'The wild goose ...'
>
> And the twelfth stroke of midnight sounded; the twelfth stroke of midnight,
> Thursday, the eleventh of October, Nineteen hundred and Twenty Eight. (226–8)

Unlike the two previous great transitions in her recorded life, Orlando does not
need to fall into a transcendent sleep to grow. The seven-day fainting-sleep is
replaced with an immediate embrace of immanence, an "oceanic feeling" in which
Orlando swims. Here the pool of the mind, the Serpentine, the Atlantic, sea and
sky and the entire world of time and space are connected.

In this biography/history/fiction, time and space signify beyond individual lives and beyond the human; with the oak-tree-scale of time, Woolf's "serious joke" seems very much to take history seriously—as the story of the English in a context broader than that of human history. Structurally, the wild goose occupies the same space as Orlando's manuscript, "The Oak Tree." It is a present absence of meaning. Yet, the wild goose *itself* (rather than the phrase which signifies fruitless pursuits, such as the Victorian pursuit of biography) is another instance of the world beyond the referent, the "green in nature" that signifies beyond letters. Woolf revised the manuscript version of this last scene in novel, which set up an equation (a static relation) between the meaning of life and the human pursuit of it,[34] to suggest this broader sense of meaning as something within *and* beyond the human ken. In this way, the specter of biology haunting biography, feminism, and history (the body, the more-than-human world), like the many selves floating inside Orlando, becomes something joyfully, materially mysterious rather than a threat to be conquered. Meaning becomes something that cannot be positively fixed, something alive and changing, a negative dialectics as ecological, material continuity. It is a relationship between place and self altogether different from the depiction of travel analyzed in the next chapter, the very real problems posed by tourism in capitalism.

---

[34] Charles G. Hoffmann quotes the following passage from the manuscript (to demonstrate the continuity between the toy boat on the Serpentine and the wild goose as symbols of ecstasy):

> And as Shelmerdine leapt from the aeroplane & ran to meet her a wild goose with its neck outstretched flew above them.
> Shel! cried Orlando.
> "The [wild] goose is –
> [The secret of life] is …"

The words in brackets are run through in the manuscript (see Hoffmann's "Fact and Fantasy in *Orlando*: Virginia Woolf's Manuscript Revisions").

Chapter 5
Consuming Culture in
A Small Place and *Among Flowers*

I would like to begin this chapter by returning to a key passage in *Candide*. In chapter nineteen, Voltaire's naïve protagonist encounters a mutilated African slave in Surinam.

> "Oh my God!" Candide said to him in Dutch. "What are you doing in this horrible condition, my friend?"
>
> "I'm waiting for my master, Mr. Vanderdendur, the famous merchant," replied the Negro.
>
> "Was it Mr. Vanderdendur," asked Candide, "who treated you this way?"
>
> "Yes, sir," said the Negro, "it's normal. They give us one pair of linen trunks twice a year as our only clothing. When we work in the sugar mills and catch our fingers in the grinder, they cut off our hand. When we try to escape, they cut off our leg. I've had both punishments. It is at this price that you eat sugar in Europe." (83)

This passage sets the stage for *Candide*'s eventual rejection of optimism as solipsistic. Sugar is not sweet when we understand how and at what human price it comes to the table. Again, perhaps the most horrifying two words of this chapter are the slave's appraisal of his situation: "it's normal." Indeed, such conditions were not only routine but pervasive; in *The Making of New World Slavery*, Robin Blackburn notes that on the Caribbean plantations, "Accidents were not uncommon when desperately tired slaves fed cane to the mill or poured the boiling cane juice from larger to smaller vats until sugar could be 'struck' from the syrup. Cane feeders could lose an arm, or worse, while a splash of boiling syrup could maim or kill" (423). By the time of *Candide*'s publication in 1759, "sugar overtook grain as the most valuable single commodity entering world trade," and most of the sugar sold in Europe came not from South America but from the "sugar islands," British and French Caribbean colonies.[1] Helen Scott emphasizes their global importance, "during the eighteenth century the sugar islands of which Antigua was one were arguably at the center of the world" (63–4). As Scott's comment suggests, Antigua

---

[1]   Blackburn calculates that the "British and French colonies supplied 70 percent of all sugar entering the Atlantic market in 1740, rising to 80 percent by 1787" (403).

was particularly important; for a time, it was the leading producer of sugar in the British Islands.[2]

Over two hundred years later Lucy, a young woman from Antigua[3] working in the United States (the title character of Jamaica Kincaid's Bildungsroman *Lucy*), contemplates the connection between sugar, freedom, and slavery:

> Paul had wanted to show me an old mansion in ruins, formerly the home of a man who had made a great deal of money in the part of the world that I was from, in the sugar industry. ... As we drove along, Paul spoke of the great explorers who had crossed great seas, not only to find riches, he said, but to feel free, and this search for freedom was part of the whole human situation. Until that moment I had no idea that he had such a hobby —freedom. Along the side of the road were dead animals—deer, raccoons, badgers, squirrels—that had been trying to get from one side to the other when fast-moving cars put a stop to them. I tried to put a light note in my voice as I said, "On their way to freedom, some people find riches, some people find death," but I did not succeed. (129)

Like the dead animals, "the origin of my [Lucy's] presence on the island—my ancestral history—was the result of a foul deed" (135), the result of people travelling to feel "free." In light of this, how do we read Lucy's attempt to "put a light note" in her voice? Is this a self-protective "making light" of the heavy thought of generations of dead slaves (performing or on their way to performing their masters' freedom) through a comparison with "mere" animals, with "roadkill"? Or is Lucy, in fact, taking this weight very seriously precisely by making this *seemingly* light ethical comparison, invoking and reversing (these dead animals are "people") the discourse of animality[4] at work in colonialism and its legacy? She tells Mariah, her white, upper-middle class employer, the story of her ride to the sugar-baron's ruined mansion in response to Mariah's feeling of freedom excited by her impending divorce. Lucy thinks, "I meant to tell her not to bank on this 'free' feeling, that it would vanish like a magic trick; but instead I told her of a ride I had taken to the country with Paul that afternoon" (128). This "free" feeling, the liberal humanist idea of freedom, like the emancipation of slaves and decolonization, is not at all what it seems. As we will see in a moment, Kincaid writes in *A Small Place* that the slaves have been freed, but *only* "in a kind of way" (80); this is why Lucy says "if he [the sugar-baron] hadn't already been dead I would have wished him so" (129). As Kincaid demonstrates here and elsewhere, postcolonial reality, for many, if not all, of the formerly colonized at home or

---

[2]   See Barbara Edlmair's *Rewriting History* (63), which cites David Barry Gaspar's "Sugar Cultivation and Slave Life in Antigua Before 1800."

[3]   While the novel only states the West Indies, Lucy tells us, "I was born on an island, a very small island, twelve miles long and eight miles wide ..." (134).

[4]   My "discourse of animality" is not so different from Cary Wolfe's "discourse of species" in *Animal Rites* (2003).

abroad, is less "post" than the term implies—perhaps, as a certain critical usage of the term suggests, only parenthetically so.

While most readers associate Kincaid with her Bildungsromane, particularly *Annie John* (1983) and *Lucy* (1990), this chapter will focus on her polemical travel narrative *A Small Place* (1988)—a radical narrative about the cost of travel— as an inverted (or obverted, negative) form of the genre, and its reverberations in her later, more "traditional" travel narrative *Among Flowers* (published by National Geographic in 2005), to explore connections between animalization, dehumanization, and exploitation; and between colonization, globalization, and environmental imperialism. An engagement with Kincaid's work, including her "garden writing," will then frame a consideration of the Bildungsroman as a multivalent discourse of origins (of species, empathy, knowledge, and the garden) in the conclusion of the book.

While many parts of the world continue to pay the human price of sugar (and of the postcolonial commodities enjoyed by those who can afford them), arguably no other part of the world "has been more radically altered in terms of human and botanic migration, transplantation, and settlement than the Caribbean" (DeLoughrey et al. 1). In their introduction to *Caribbean Literature and the Environment*, the first ecocritical[5] collection on Caribbean literature, the editors ground the volume in the crucial assertion that "natural histories," including what counts as natural, "are deeply embedded in the world historical process" (which is, as we shall see, particularly important to Kincaid) and that Caribbean landscapes "continue to be misunderstood for reasons that can be traced to early Caribbean colonists" (6). For example, the "myth of fertility confused plant diversity with an extraordinary yield for food, leading readers [of early reports of the Caribbean, such as Columbus's journals] and many a current-day tourist to assume that one need not labor in tropical climates for sustenance" (6). For this and other reasons, including racist phobias generated by an increase of forcibly transplanted Africans in the American tropics, at "the height of plantation slavery, Europeans began to separate 'culture' from its epistemological root, 'cultivation,' and attribute degeneracy to those involved in tropical agriculture" (7). These contradictory

---

⁵   Though an ecocritical project, this volume targets what might be construed as the unconscious racism of some earlier works in the field: "Although North American ecocritics often inscribe an idealized natural landscape that is devoid of human history and labor, the colonization and forced relocation of Caribbean subjects preclude that luxury and beg the question of what might be considered a natural landscape. Against the popular grain of U.S. ecocritical studies, we argue that addressing the historical and racial violence of the Caribbean is integral to understanding literary representations of its geography" (2). In a similar vein, Susie O'Brien has commented on a corollary problem, the way in which certain works of ecocriticism, in an effort to become "more worldly" (multicultural, transcultural) risk "repeating the imperialist rhetoric it [ecocriticism] purports to undo. ... the various routes that ecocriticism proposes for 'getting back to the world' might, in their haste to overleap the boundaries of culture—discursive, national, and institutional—lead merely to a more globalist form of parochialism" ("The Garden and the World" 169).

myths and attitudes are at play in eighteenth- and nineteenth-century depictions of sugar plantations as "gardens." As Melanie Murray demonstrates, "when economically productive, the plantation is depicted as a paradisal garden. When not prosperous, it becomes a wilderness in need of taming, implying a justification for conquest" (63). This imperial exercise of "mastery" shapes, as Mimi Sheller suggests, "the ways in which later tourists came to gaze upon the landscape and experience bodily the pleasures of Caribbean travel" (26).

Jamaica Kincaid's narrative essay *A Small Place* recounts and contrasts her experience growing up in Antigua with Western and Antiguan, tourist and native, experiences of the island today, including her own visit to the island after living in America for over two decades. "Discovered" by Columbus in 1493, Antigua was colonized by the British in 1632, who also claimed the nearby island of Barbuda in 1678, both of which became part of the British Leeward Islands colony (defederated in 1956). Only in 1981 did Antigua and Barbuda gain full political independence. This independence, however, is more theoretical than actual; again, Kincaid tells us, "Eventually, the masters left, in a kind of way; eventually, the slaves were freed, in a kind of way" (80).

No longer a mass of sugar plantations, postcolonial Antigua's main export is the memory of a tropical "paradise" vacation. Every year, hundreds of thousands of tourists in search of "affordable" color, warmth, and bodily pleasure, flock from Britain, the United States, and other rich countries to Antigua and Barbuda, where the pound, euro, or dollar goes far, the sun is constant and bright, and the "natives" are advertised as welcoming.[6] The official visitor's guide of the Antigua and Barbuda Department of Tourism[7] lists twenty-five reasons to visit Antigua, including, at number seven, "It's an English-speaking island, so there are no communication barriers." If that isn't convincing (and Kincaid demonstrates it shouldn't be) the guide attempts to reassure with number fourteen: "the warmth and hospitality of Antiguans and Barbudans, long considered some of the friendliest people on the planet!" The visitor's guide literally banks on the potency of what Sheller terms "the myth of 'easy' living in the Caribbean ... accompanied by assumptions that the 'natives' were available to serve tourists, offer friendly greetings, and become part of the scenery" (29).

Kincaid begins *A Small Place* describing to "you," the reader, your experience of Antigua. In this narrative position of the tourist, readers literally come into Antigua consuming it, economically and environmentally. As we "arrive" in Antigua, Kincaid's narrative voice becomes the voice in our heads. To demonstrate

---

6   Two decades after the publication of *A Small Place*, the islands received 220,793 tourists (visitors declared to be on vacation), according to the official corporate website of the Ministry of Tourism of Antigua and Barbuda: http://www.tourismantiguabarbuda.gov. ag/tourism_programs/pdf/statistics/stay_over_arrivals_by_purpose_of_visit_2008.pdf.

7   From the Encyclopedia Britannica entry (http://www.britannica.com/EBchecked/topic/916749/Antigua) follow the link to the "Official Site of the Antigua and Barbuda Department of Tourism."

the way in which this reading experience feels both immediate and conspicuously mediated, I will quote the opening at length:

> If you go to Antigua as a tourist, this is what you will see. If you come by aeroplane, you will land at the V.C. Bird International Airport. Vere Cornwall (V.C.) Bird is the Prime Minister of Antigua. You may be the sort of tourist who would wonder why a Prime Minister would want an airport named after him— why not a school, why not a hospital, why not some great public monument? You are a tourist and you have not yet seen a school in Antigua, you have not yet seen the hospital in Antigua, you have not yet seen a public monument in Antigua. As your plane descends to land, you might say, What a beautiful island Antigua is—more beautiful than any of the other islands you have seen, and they were very beautiful in their way, but they were much too green, much too lush with vegetation, which indicated to you, the tourist, that they got quite a bit of rainfall, and rain is the very thing that you, just now, do not want, for you are thinking of the hard and cold and dark and long days you spent working in North America (or, worse, Europe), earning some money so that you could stay in this place (Antigua) where the sun always shines … since you are a tourist, the thought of what it might be like for someone who had to live day in, day out in a place that suffers constantly from drought, and so has to watch carefully every drop of fresh water used (while at the same time surrounded by a sea and an ocean—the Caribbean Sea on one side, the Atlantic Ocean on the other), must never cross your mind. (3–4)

Oddly intimate and omniscient, the narrator is there before us (has seen the school, the hospital), knows the island's geography and climate, and moves above it and in and out of the rhetorical reader's head. But, as the narrative progresses, oppositional parenthetical asides accumulate, calling attention to textual, material, and historical mediation, suggesting a specific, embodied voice. Scott notes the particular ontological overlap of the narrator and tourist: "While the narrator addresses a very specific persona—the hypothetical tourist—she at the same time is describing what *she* sees as she approaches [as Kincaid returns to] the island" (70). And while Barbara Edlmair suggests that Kincaid's ambiguous narrative and subject position as an American in Antigua remains "unproblematized" (79), Isabel Hoving argues almost the reverse, that although *A Small Place* creates a "remarkably unshakeable" opposition of "I" and "you," these pronouns display an intentional ambiguity: "The 'I' identifies herself first as an inhabitant of the island Antigua … second, as a migrant who returned temporarily to her motherland but has obtained a different perspective on the place where she was born, and third, a migrant who sometimes identifies herself as one among millions of victims of slavery and colonization" (*In Praise* 193–4). Though Hoving sees these pronouns operating in "different historical and political frameworks," creating "the impossibility of a straightforward 'we'" (202), the "unshakeable" quality of this narrative structure suggests that these differences are intimately connected, both textually and materially, a *continuity* that insists on the neocolonial politics of postcolonial Antigua. Indeed, throughout the text, tone and narrative perspective shift between the rhetorical reader's consciousness and the narrator's, between "main" text and parenthetical aside, voicing the unsustainable dichotomy of tourist and native, colonizer and colonized, master and slave, as a continuous historical process created by colonization and perpetuated by neoliberal globalization.[8]

As we will see in a moment, the anger of *A Small Place* is certainly insistent. Kincaid told Donna Perry in an interview that *The New York Times* described *A Small Place* as lacking the "charm" of her earlier work:

> Really, when people say you're charming you are in deep trouble. I realized in writing that book that the first step to claiming yourself is anger. You get mad. And you can't do anything before you get angry. ... I can see that *At the Bottom of the River* was, for instance, a very unangry, decent, civilized book and it represents sort of this successful attempt by English people to make their version of a human being or a person out of me. (Perry 133)

Unlike most reviewers, many critics applaud the anger in *A Small Place*, particularly when it is directed at the tourist (predominantly white, middle or upper class Europeans and Americans), but this is not always the case. Scott, for example, questions the character of other moments of anger in the text, "the distinct separation between speaker and Antiguans (West Indians), and the attitude of scorn and disgust expressed by the former for the latter" (68). Allison Donnell, on the other hand, reads this move as a positive refusal of the "soft target of the tourist in favour of a range of more contentious offenders" ("She Ties" 43), and Hoving argues the text's anger "exceeds its irony," eschewing realism in order to critique postcolonial nationalism (203).

Clearly, there is a good deal of disagreement about *A Small Place*, about the function of its narrative structure, its tone, intent, and effect. Indeed, the text's ability to polarize (even a community of relatively like-minded critics) makes sense given that its energy is polemical, focused on Caribbean tourism, its roots in slavery and colonial rule, *and* the people and institutions it supports or fails to support in postcolonial Antigua (gambling, prostitution, and widespread corruption on the one hand, and schools, libraries, hospitals, and social services on the other). *A Small Place* presents an intentionally uncomfortable narrative

---

[8]   And this dichotomy continues to pose problems throughout the text. Moira Ferguson contends it "presents a point of view that demands some mediation" (*Colonialism and Gender* 132), while Caren Kaplan suggests that the text itself "mediates the very oppositions it constructs, breaks open contradictions, makes connections" (28). Allison Donnell sees *A Small Place* as "a consummate work of ventriloquism that deploys a whole series of voices in order to debate the values and limitations of the cultural discourses and positions associated with postcolonialism," including what Giovanna Covi calls a "faked, primitive child-like voice" (qtd. in Donnell) as well as "an astonishingly arrogant propensity to speak for others" and an "adoption of colonial discourse" ("She Ties ..." 42–3). Donnell concludes that the text "reiterates the radical possibilities of mimicry" while "dismantling cultural positions" and "the constraints of 'correctness'" that paralyze thought and action (47, 48).

position and reading experience in form and content to counter the tourist's (and tourist-reader's) demand for comfort and pleasure, because capitalism was and is so much more than uncomfortable for a great many. As Moira Ferguson has it, "Tourism is a modern version of colonialism; it is domination's new, lucrative face" (*Jamaica Kincaid* 86). It is in this context that I undertake a reading of *A Small Place* as a provocative, genre-bending Bildungsroman of capitalism, in which the fictions of capitalism are polemically engaged through the reality of life in postcolonial Antigua.

When *you*, the tourist, do finally pass one of those Antiguan institutions V.C. Bird could have named after him, you see "a building sitting in a sea of dust and you think, It's some latrines for people just passing by, but when you look again you see the building has written on it PIGOTT'S SCHOOL" (7). The narrator tells us that we tourists needn't feel "funny" about the heartbreaking poverty before our eyes, about how the poor became poor and the rich became rich in the first place: "the West got rich not from the free (free—in this case meaning got-for-nothing) and then undervalued labour, for generations, of the people like me you see walking around you in Antigua but from the ingenuity of small shopkeepers in Sheffield and Yorkshire and Lancashire, or wherever ..." (9–10).[9] Because we take our assumptions, our cultural narratives and their ideologies with us, we—the tourists—can feel comfortable *enjoying* the historical repercussions of slavery. So, "you needn't let that slightly funny feeling you have from time to time about exploitation, oppression, and domination develop into full-fledged unease, discomfort; you could ruin your holiday"; after all, as the narrator asserts in the tourist's defensive voice, "They are not responsible for what you have; you owe them nothing; in fact, you did them a big favour. And you can provide one hundred examples" (10). But in parentheses, in a space between the tourist's ideological construction of the past and "funny" feeling in the present, the narrator tells a literal home-truth, "(isn't that the last straw; for not only did we have to suffer the unspeakableness of slavery, but the satisfaction to be had from 'We made you bastards rich' is taken away too)" (10).

The bad faith of the capitalist historiography Kincaid references is the bad faith of capitalism, which relies on a discourse of animality, a narrative that enables capitalism to categorize certain humans and nonhumans as "resources" for production. As Blackburn writes, "In practice, slaves were conceived of as an inferior species, and treated as beasts of burden ... Yet like all racist ideologies, this one was riddled with bad faith. The slaves were useful to planters precisely because they were men and women capable of understanding and executing complex orders, and of intricate co-operative techniques" (12). The construction of non-Europeans as "outside of culture and morality in some 'wild' and 'natural' state" (15)[10] was an embodiment of phobic fantasies and utopian projections,

---

[9]   It is interesting that our narrator gets place "wrong" here; Sheffield is *in* Yorkshire.

[10]   For example, Blackburn notes that the word "cannibalism" was constructed from the name of the Carib people (22).

necessitating their domestication as "chattel" (15–16) to make them "safe" for domination. What Blackburn writes of emergent capitalist cultures was true of empire at its height and is true of globalization of today: the logic of capitalism "only recognize[s] the humanity of those who had something to sell" (16). Yet even this assessment seems an optimistic appraisal of capitalism's ability to value life, human or otherwise. As liberal economist Lester Thurow concluded, "'Survival of the fittest' and inequalities in purchasing power are what capitalist efficiency is all about. ... To put it in its starkest form, capitalism is perfectly compatible with slavery. Democracy is not" (qtd. in Street 20). And so it should be no surprise that "[f]ew if any aspects of contemporary capitalism are less democratic than precisely its tendency toward globalization" (Street 22).[11]

Obverting Fiction

The narrator of *A Small Place* articulates the ideology of social Darwinism at work in capitalist mythmaking. *You* the tourist can enjoy your holiday; *you* may rest assured that all is well in the world because "their ancestors were not clever in the way yours were and not ruthless in the way yours were, for then would it not be you who would be in harmony with nature and backwards in that charming way?" (17).

Here, Kincaid explicitly characterizes the bad faith of capitalism, of the tourist/colonizer's racist ideologies and instrumental reason, as the bad faith of the discourse of animality. Kincaid's essay dramatizes and targets this bad faith rhetorically and viscerally even at the sentence level. Readers literally come into a culture consuming it, only to find themselves dehumanized by this dehumanizing consumption. Structurally, *A Small Place* may be read not only as the "anti-Bildungsroman" of the tourist (who travels to learn nothing but, as the reader, learns a great deal) but also of an entire people pressed into Western culture and global economy, reversing the Western genre of the individual who learns "his place" through travel by narrating the story of an entire people "put in their place," people for whom "development" means being "too poor to go anywhere. Too poor to escape the reality of their lives; and they are too poor to live properly in the place they live, which is the very place you, the tourist, want to go" (19). By turning her attention to the subject(s) of Antigua, Kincaid casts critical travel writing as the counterpart of the liberal narrative of travel and acculturation.[12] Indeed, *A Small Place* obverts the reversal of tourism itself, of the way in which the tourist turns everyday "banality and boredom" into pleasure: "when the natives see you, the

---

[11]   Paul Street charts the reasons why American CEOs, and transnational corporations generally, prefer to operate in regimes based on minority-rule: low wages, few if any environmental and labor regulations, and similar assumptions about decision making processes and human rights.

[12]   By obverting the genre of the Bildungsroman, Kincaid quite literally puts a face on the coin of capital.

tourist, they envy you, they envy your ability to leave your own banality and boredom, they envy your ability to turn their banality and boredom into a source of pleasure for yourself" (19).

In the tradition of the Bildungsroman as a story of education that educates (as we see in *Candide, Frankenstein,* and *Orlando*), *A Small Place* teaches us why the ignorant tourist is ugly, why the native is angry, why the native seems to the tourist "in harmony with nature and backwards in that charming way." From the Pigott's School, easily mistaken for a latrine, we see that the library damaged in "The Earthquake" of 1974 remains unrepaired more than a decade later (8–9). Thankfully, *you* (the tourist) have brought your books with you, including "one of those new books about economic history, one of those books explaining how the West (meaning Europe and North America after its conquest and settlement by Europeans) got rich" (9). This book tells capitalism's story of itself, its own origin-story, which, *again*, goes something like this: the West got rich

> not from the free (free—in this case meaning got-for-nothing) and then undervalued labour, for generations, of the people like me you see walking around you in Antigua but from the ingenuity of small shopkeepers in Sheffield and Yorkshire and Lancashire, or wherever; and what a great part the invention of the wristwatch played in it, for there was nothing noble-minded men could not do when they discovered they could slap time on their wrists just like that … (10)

This book, which *you* have brought all the way from home, is one of capitalism's Bildungsromane. With *A Small Place*, Kincaid gives us, instead, a Bildungsroman of capitalism: the story of not only where but who capital came from, where they are now, and why.

The ugliness of the tourist is the acculturation, the Bildungsroman that you bring with you as you "make a leap from being that nice blob just sitting like a boob in your amniotic sac of the modern experience to being a person visiting heaps of death and ruin and feeling alive and inspired at the sight of it"—a "leap" that seems no leap at all—to being "a person marveling at the harmony (ordinarily, what you would say is the backwardness) and the union these other people (and they are other people) have with nature" (16). This is why the narrator pronounces: "An ugly thing, this is what you are when you become a tourist, an ugly, empty thing, a stupid thing, a piece of rubbish pausing here and there to gaze at this and that, and it will never occur to you that the people who inhabit the place in which you have just paused cannot stand you" (17). *Your* "ugliness" is not simply or only an aesthetic and cultural incongruity—your "displacedness," a feeling of which you already have at home—but also your solipsism, your lack of historical consciousness, social or ecological:

> That water—have you ever seen anything like? Far out, to the horizon, the colour of the water is navy-blue; nearer, the water is the colour of the North American sky. From there to the shore, the water is pale, silvery, clear, so clear that you

can see its pinkish-white sand bottom. Oh, what beauty! Oh, what beauty! You
have never seen anything like this. ... Still standing, looking out the window,
you see yourself lying on the beach, enjoying the amazing sun (... a sun that
is your personal friend). You see yourself taking a walk on that beach, you see
yourself meeting new people (only they are new in a very limited way, for they
are people just like you). You see yourself eating some delicious, locally grown
food. You see yourself, you see yourself. You must not wonder what exactly
happened to the contents of your lavatory when you flushed it. ... the contents
of your lavatory might, just might, graze gently against your ankle as you wade
carefree in the water, for you see, in Antigua, there is no proper sewage-disposal
system. (12–14)

This is not mere metaphor but it is also metaphor: the literal and metaphorical
shit of the West "returns" to its origin (as the human "rubbish"—the wealth of
the west—returns to the truth of its origins). The solipsism of the tourist is grazed
by the historical, material reality of his body and its place in the socioecological
world. The present interconnectedness of body, society, and environment (from
the connection between rainfall, tourism, and corruption, to the more recently
reported connection between tourism, human sewage, and the rapid decline of
elkhorn coral in the Caribbean[13]) parallels the connections between the present
and past of Antigua.

And so, the school *you*, the tourist, mistook for a lavatory is in a sense both
repetition and consequence of the enslavement of generations of peoples. The
tourist's solipsism is the very "amniotic sac of modern experience" he sought to
leave at home: his lack of historical (social) consciousness is part of his lack of
ecological consciousness and vice-versa. The narrator continues:

But the Caribbean Sea is very big and the Atlantic Ocean is even bigger; it would
amaze you to know the number of black slaves this ocean has swallowed up.
When you sit down to eat your delicious meal, it's better that you don't know
that most of what you are eating came off a plane from Miami. And before it got
on a plane in Miami, who knows where it came from? A good chance it came
from a place like Antigua first, where it was grown dirt-cheap, went to Miami,

---

[13]   National Public Radio reported on August 17, 2011: "Human beings occasionally
get diseases from animals, such as swine flu, rabies and anthrax. A new study finds that
humans can also spread disease to wildlife, with grim results. A bacterium from our guts
is now rampaging through coral reefs in the Caribbean. Those reefs were already in slow
decline, but they took a huge hit starting in 1996, when a disease called white pox appeared
in the Florida Keys. 'Since that time, elkhorn coral—the species it affects—has declined 88
percent in the Florida Keys,' says Kathryn Sutherland, a reef ecologist at Rollins College
in Florida. ... 'But this is a problem Caribbean-wide,' Sutherland says, 'and there's a
widespread lack of wastewater treatment in the wider Caribbean region.' And that's bad news
for the elkhorn coral. Due largely to the disease spreading from humans, it's been tagged
as vulnerable on the endangered species list." http://www.npr.org/2011/08/17/139705482/
caribbean-coral-catch-disease-from-sewage.

and came back. There is a world of something in this, but I can't go into it right now. (14)

The "world of something in this" is nothing less than the world itself, the historical, material interconnectedness of social and ecological worlds, human and more than human, and the interconnected forms of domination at work in global capitalism (and its international trade policies—as the narrator suggests later in the text, the rule of "Gross National Product"). The floating memory of dead slaves and tourists' shit: there is a *large* world of something in this, of racist identifications, capitalism, and dehumanization (which will surface, again, in this chapter and the conclusion).

The same system that constructs the tourist's lack of historical consciousness, his false consciousness, the social Darwinism of capitalism's ugly origin story ("their ancestors were not clever in the way yours were") created and continues a parallel problem in Antigua; it is the reason why some people in this place cannot "see themselves in a larger picture," in politics and history (52). The legacy of colonization in the Caribbean is the rule of the GNP, the economic "restructuring" and other forms of corruption both foreign and domestic that perpetuate a cyclical lack of resources, opportunity, and equity.

After describing Pigott's school, the narrator recounts a telling story about the headmistress of a girls' school, placed by the colonial office: "This woman was twenty-six years old, not too long out of university, from Northern Ireland, and she told these girls to stop behaving as if they were monkeys just out of trees. No one ever dreamed that the word for any of this was racism" (29). This critique of Antiguan consciousness is not, as some suggest, an attack on Antiguans, and Kincaid includes herself in this critique: "*Our* perception of this Antigua ... was not a *political* perception. The English were ill-mannered, not racists; the school headmistress was especially ill-mannered, not a racist; the doctor was crazy ... he was also not a racist; the people at the Mill Reef Club were puzzling ... not racists" (34, italics mine). And then there is the Hotel Training School:

> They speak of emancipation as if it happened just the other day, not over one hundred and fifty years ago. ... And perhaps there is something in that, for an institution that is often celebrated in Antigua is the Hotel Training School, a school that teaches Antiguans how to be good servants, how to be a good nobody, which is what a servant is. In Antigua, people cannot see a relationship between their obsession with slavery and emancipation and their celebration of the Hotel Training School (graduation ceremonies are broadcast on radio and television) ... (55)

Kincaid writes that Antiguans "cannot see that they might be part of a chain of something, anything" and so they experience an event "as if it were sitting on top of their heads" (53). This is why they do not see the connections between slavery and servitude, between being called a "monkey" and being trained to be a "nobody." Kincaid's colonial education in Antigua taught her to memorize Wordsworth and view England as the "motherland"; it did not teach her to question the processes and values of this education or the other tools of colonial rule. She had to learn this on her own, first by getting angry, by recognizing that, as she asserts repeatedly in the text, the most serious "crime" has been committed against her and millions of others. And this is why our narrator becomes so very "angry ... to hear people from North America tell me how much they love England, how beautiful England is, with its traditions. ... But what I see is the millions of people, of whom I am just one, made orphans: no motherland ... no tongue" (31). While this "anger ... stays unsettlingly close to colonial discourse" (Hoving 194), Kincaid's discussion of Antiguan consciousness is not (or not only) an angry mimicry of colonial discourse but an explicit critique of the legacy of the enslavement and colonization that made Antiguans orphans in time as well as space; colonization disrupted the sense of history and self engendered by rootedness in place, language, and culture.

For this reason, our narrator asks us, directly, a series of questions that the tourist cannot adequately answer:

> Do you ever wonder why some people blow things up? I can imagine that if my life had taken a certain turn, there would be the Barclays Bank, and there I would be, both of us in ashes. Do you ever try to understand why people like me cannot get over the past, cannot forgive and cannot forget? There is the Barclays Bank. The Barclay brothers are dead. The human beings they traded, the human beings who to them were only commodities, are dead. (26)

People die but their institutions remain, and in this way slavery is still alive in present Antigua (and, as we saw in *Lucy*, to varying degrees, everywhere touched by it). This is the present tense of narration, the present reality of native and tourist (here, criminal) alike, presented for a moment as the alternate-life of our narrator: "And when I blow things up and make life generally unlivable for the criminal (is my life not unlivable, too?), the criminal is shocked, surprised. But nothing can erase my rage ... for this wrong can never be made right ..." (32). The criminal is "surprised" but not *really* surprised; the questions the tourist cannot answer are truly, in fact, questions he *will not* answer—cannot answer and maintain, to any degree, the cultural fantasies of capitalism. The reality of the native, the "terrorist," the subaltern, the hungry, the poor—the other in, as I will argue in a moment, various forms—is the latent knowledge of the tourist, of capitalism. Hoving astutely observes with respect to the tourist's ugliness (which I have read as a manifestation of capitalism's ugly origin-story): "he has always suspected it. There is no gap between what the native knows and what the tourist (really) knows, or should know. ... The tourist apparently has to suppress certain insights to maintain his position as a privileged tourist. This blindness characterizes the tourist also at those moments when he is not a tourist, but at home" (214).

And so, our narrator asks *us*, "Do you know why people like me are shy about being capitalists? Well, it's because we, for as long as we have known you, *were* capital, like bales of cotton and sacks of sugar, and you were the commanding, cruel capitalists, and the memory of this is so strong, the experience so recent, that we can't quite bring ourselves to embrace this idea that you think so much of" (36–7, italics in original). As Marx writes in Volume One of *Capital*, "It is in tropical culture, where annual profits often equal the whole capital of plantations, that negro life is most recklessly sacrificed. It is the agriculture of the West Indies, which has been for centuries prolific of fabulous wealth, that has engulfed millions of the African race" (377). "We," the colonist-tourist, "first-world" beneficiary of imperialism and globalization, are accused of this historical and present violence, for (as Marx maintains in *Capital* and elsewhere) *nothing* in capitalism is neutral, particularly—as Kincaid demonstrates—what (capitalism) claims to be neutral: "Switzerland is a neutral country, money is a neutral commodity, and time is neutral, too" (60). Read as a Bildungsroman of capitalism, a narrative of the latent knowledge of capitalism's own ugly origin story, *A Small Place* teaches us to *see* what we already know.

In the institution of Barclays Bank (among others), the violence of slavery became and continues as the present violence of poverty in capitalism, in neoliberal global economy: "The Barclay brothers, who started Barclays bank, were slave traders. That is how they made their money. When the English outlawed the slave trade, the Barclay brothers went into banking. … for just look at how rich they became with their banks borrowing from (through their savings) the descendants of slaves and then lending back to them" (25–6). As Marx argued with respect to capitalism generally, "*Mutato nomine de te fabula narrator.*[14] For slave trade, read labour-market" (378); for Barclays Bank, read World Bank. *Life and Debt*, Stephanie Black's 2001 documentary, uses passages from *A Small Place*, adapted by Kincaid herself, to comment on "the stories of individual Jamaicans whose strategies for survival and parameters of day-to-day existence are determined by the U.S. and other foreign economic agendas."[15] One particularly resonant story is that of the furniture and cabinet maker who, though his furniture is no longer bought, can barely keep up with the demand for coffins (interestingly, Kincaid's father was a cabinetmaker, as is Lucy's[16]).

Far from "fixing" the problems caused by colonial rule, the neoliberal economic policies of the International Monetary Fund perpetuate and exacerbate poverty and violence. The IMF agenda of "monetary austerity, currency devaluation, and lowering wages" and the free-market "structural adjustments" of the World Bank lead to a cycle of recession and high-interest loans, the death of local economies and self-determination: "These policies are supposed to benefit Third World economies by integrating them into the global market. What actually happens is that Third World people suffer, while commercial banks in the North collect a great deal of interest. In Jamaica, only 5 percent of total money borrowed since 1977

---

[14] "The name is changed, but the tale is told of you!" (Horace, *Satires*, Book 1, Satire 1).
[15] http://lifeanddebt.org/about.html.
[16] *Conversations with American Novelists* 27.

has been able to stay inside the country."[17] Scott's assessment of globalization is, I think, more precise, in that she does not assume that these policies are (or were ever) supposed to benefit anyone other than those already in power:

> Neoliberal globalization has economic and geopolitical motivations (strategic positioning vis-à-vis other powers; control over profits from oil refining, mineral extraction, offshore manufacturing, tourism, cheap labor); takes economic, political, and cultural forms; and is ultimately backed by *the threat or reality of military force*. It thus conforms to the definition of imperialism. (2, italics mine)

This is where the violence of *A Small Place* comes from; it is a response to the violence of capitalism, in its many forms. It is why "people like me ... mutilated [the] bodies of you, your wife, and your children" in the bungalow of your plantation, why "you say to me, 'Well, I wash my hands of all of you,' and you leave and from afar you watch as we do to ourselves the very things you used to do to us" and "forget your part in the whole setup, that bureaucracy is one of your inventions, that Gross National Product is one of your inventions, and all the laws that you know mysteriously favor you" (35–6). Of course, *you* have never left and do not watch, disinterestedly, from afar. This is why the tourist visits "heaps of death and ruin" in the Caribbean and why some people (some natives) blow things up.

I'll Be a Monkey's Uncle

Our narrator proclaims that the violence of colonialism was so horrific and persists in such a way that the English should be perpetually mindful of "the irrevocableness of their bad deeds, for no *natural* disaster imaginable could equal the harm they did" (24, italics mine). While capitalism, its colonial and neocolonial phases, is an idea ("this idea that you think so much of"), it is one which transforms material and social relations, creates and negates subjectivity ("we ... *were* capital, like bales of cotton and sacks of sugar ...").[18] Capitalism

---

[17]   http://lifeanddebt.org/about.html. Scott summarizes the history well: "A response to the crises in global capitalism that began early in the 1970s, and implemented by the main global powers and the multilateral financial institutions they govern, most importantly the World Bank and the International Monetary Fund (IMF), neoliberal globalization rests on economic deregulation, financialization, and privatization, in the name of the 'free market' and 'free trade.' From the 1970s 'Third World' nations were encouraged to take out loans from foreign governments, private banks, and international institutions. With the debt crisis of 1982 many indebted countries were forced to carry out Structural Adjustment Programs (SAPs) in return for loan assistance from the World Bank or IMF. SAPs involve devaluation of currency; an orientation on exports and foreign capital investment; liberalization of financial markets; and cuts in government spending, subsidies, and wages" (2).

[18]   However, in a small place (and elsewhere in capitalism) ideas seem detached from material reality. "And might not knowing why they are the way they are," our narrator asks,

turns "others" into a "resource" for production, other living beings into objects (in fact, it turns everyone, to varying degrees, into a resource for the cycle of production/consumption). And this is why our narrator states: "*Even* if I really came from people who were living like monkeys in trees, it was better to be that than what happened to me, what I became after I met you" (37, italics mine). Better to *even* be of a people living "like monkeys in trees" than living objects for sale; however, monkeys have long been and still are *objects* for sale—indeed, one might say animals are the "original" objects for sale (the commodities of meat, labor, spectacle, and knowledge, on farms and in factories, zoos, and labs).

In these passages, the text seems to do two things at once. It acknowledges the interconnectedness of social and ecological worlds—the destruction of people and land—while reinforcing the separateness of (its ideas of) nature and culture—natural disaster and colonialism, monkeys and human beings (a tension that, as we will see in a moment, becomes more apparent in *Among Flowers*). The *even* of "Even if I really came from people who were living like monkeys in trees" duplicates (if only to foreground) the bad faith of the colonizer's racism. Let's return for a moment to a passage from Blackburn's *The Making of New World Slavery*, quoted earlier in this chapter: "In practice, slaves were conceived of as an inferior species, and treated as beasts of burden ... Yet like all racist ideologies, this one was riddled with bad faith. The slaves were useful to planters precisely because they were men and women" (12). The bad faith of the slaveholder is readily apparent. However, Blackburn's analysis begs the question often termed "the question of the animal": are we really completely different from what the West calls "beasts"? As the Introduction of this book puts forth, a good deal of scientific research—and less formal and even latent knowledge—suggests that we aren't.

While Social Darwinism is a gross misrepresentation of Darwinian thought,[19] the discourse of animality at work in capitalism is both a product and process of anthropodenial, an anti-evolutionary ideology. In this passage from *The Ape and the Sushi Master*, quoted in the Introduction, de Waal argues, "if anthropomorphism carries a risk, its opposite carries a risk too. To give it a name, I propose *anthropodenial* for the a priori rejection of shared characteristics between humans and animals when in fact they may exist" (68–9).[20] A Darwinian thinker, de Waal emphasizes evolution as a continuous process in both time and space and a

---

"why they do the things they do, why they live the way they live and in the place they live, why the things that happened to them happened, lead these people to a different relationship with the world, a more demanding relationship, a relationship in which they are not victims all the time of every bad idea that flits across the mind of the world?" (56–7).

[19] Social Darwinism is "like a secret mistress ... rejected as soon as the daylight shines on real Darwinism. This ideology was unleashed by British political philosopher Herbert Spencer, who in the nineteenth century translated the laws of nature into business language, coining the phrase 'survival of the fittest' (often incorrectly attributed to Darwin)" (*Age* 28).

[20] "Ever since Descartes, the air has been filled with warnings against anthropomorphism. ... But getting rid of anthropomorphism is neither easy nor risk free.

continuity between the species, particularly between human and other animals. He suggests that the West's "historic lack of exposure to monkeys and apes has only reinforced its sense of human uniqueness" (35), while the "presence of monkeys in India, China, Japan—in contrast to the Middle East and Europe—may have strengthened people's closeness to nature: seeing other primates makes it hard for us to deny that we are a part of nature" (190). This lack of exposure has not, however, stopped the West from making "monkeys" the special tools for meaning and objects of knowledge; in fact, this lack of exposure seems, in part, to have produced it. As Haraway argues in *Primate Visions*, "Monkeys and apes have a privileged relation to nature and culture for western people: simians occupy the border zones between those potent mythic poles" (1). Just as Said demonstrated that the "Orient has been a troubling resource for the production of the Occident" (10), so too the nonhuman primate came to stand in for the all-important line between nature and culture: "here, the scene of origins is not the cradle of civilization, but the cradle of culture, of human being as distinct from animal existence. ... Simian orientalism means that western primatology has been about the construction of the self from the raw material of the other, the appropriation of nature in the production of culture, the ripening of the human from the soil of the animal" (10–11). As a Western origin story of culture, this is also an origin story of whiteness, masculinity, and mind[21] and, in a sense, a discourse about the origin of origins, sexuality: "Traditionally associated with lewd meanings, sexual lust, and the unrestrained body, monkeys and apes mirror humans in a complex play of distortions over centuries of western commentaries on these troubling doubles" (11).[22]

In terms almost identical to the racist remark made by the headmistress described in *A Small Place* (herself a colonial subject[23]), Annie of *Annie John* (Kincaid's first Bildungsroman) nearly reprimands her estranged friend, Gwen, who had "degenerated into complete silliness" with "schoolgirl traits she did not have when she was actually a schoolgirl." "When we were saying our goodbyes, it was all I could do not to say cruelly, 'Why are you behaving like such a monkey?'" (137). While *Annie John* tells the story of a young girl growing up in Antigua from ten to seventeen (at which point she leaves the island to study in England),

---

By changing our language as soon as we describe animals, we may be concealing genuine similarities" (35).

[21]   The construction of "the clarity of white from the obscurity of color, the issue of man from the body of woman, the elaboration of gender from the resource of sex, the emergence of mind by the activation of body" (11).

[22]   Discussed in the Introduction, a significant proportion of new work in primatology, specifically in cultural biology, is quite different. More broadly, the concept of trans-species research argues against treating other animals as a resource for the production of the human (and for human production).

[23]   While the Irish Free State was formed by treaty in 1921 (23 southern counties and 3 counties in Ulster) and became an independent republic in 1949, the other counties of Ulster formed Northern Ireland.

Lucy, quoted earlier in this chapter, chronicles the life of a strikingly similar girl, now nineteen, who comes to America from Antigua (again, to work as an au pair for an upper middle class family, as Kincaid herself did).[24] Annie's unasked question reveals the internalized rhetoric of the colonizer, a paradoxical demand for mimicry[25] (*don't behave like a monkey; act like—"ape"—me!*) through the speciesist discourse of the monkey, a discourse continued in *Lucy*.

While *"Annie John* foretells the mature, radical politic of *A Small Place*. At one level *A Small Place* is *Annie John,* part 2" (Ferguson, *Colonialism ...* 133), *Lucy* may similarly be read as a working through of *Annie John* and *A Small Place* (or, *A Small Place*, part 2). In an early scene of the novel, Lucy's employers Mariah and Lewis comment on her separateness; their seeming concern positions her as wholly alien, a member of another species, a visitor from another planet of experience (without actually recognizing any real, specific cultural differences):

> It was at dinner one night not long after I began to live with them that they began to call me Visitor. They said I seemed not to be a part of things, as if I didn't live in their house with them ... Lewis looked at me, concern on his face. He said, 'Poor Visitor, poor Visitor,' over and over, a sympathetic tone in his voice, and then he told me a story about an uncle he had who had gone to Canada and raised monkeys, and of how after a while the uncle loved monkeys so much and was so used to being around them that he found *actual* human beings hard to take. He had told me this story about his uncle before, and while he was telling it to me this time I was remembering a dream I had had about them: Lewis was chasing me around the house. I wasn't wearing any clothes. The ground on which I was running was yellow, as if it had been paved with cornmeal. Lewis was chasing me around and around the house, and though he came close he could never catch up with me. Mariah stood at the open windows saying, Catch her, Lewis, catch her. Eventually I fell down a hole, at the bottom of which were some silver and blue snakes.
>
> When Lewis finished telling his story, I told them my dream. ... Their two yellow heads swam toward each other and, in unison, bobbed up and down. Lewis made a clucking noise, then said, Poor, poor Visitor. And Mariah said, Dr. Freud for Visitor, and I wondered why she said that, for I did not know who Dr. Freud was. Then they laughed in a soft, kind way. (14–15, italics mine)

Racist comparisons of humans to nonhuman animals are not unusual in colonial and postcolonial literatures. What is particularly interesting about Lewis's story (and the fact of his repetition of the story) of his uncle's "regression" is its function

---

24   Numerous critics have remarked that Kincaid's work is autobiographical, and Kincaid seems to play with the notion of autobiography, in these texts and *Autobiography of My Mother*, 1996, and *Mr. Potter*, 2002.

25   Following Homi K. Bhabha's theory of mimicry in *The Location of Culture*: "colonial mimicry is the desire for a reformed, recognizable Other, *as a subject of difference that is almost the same, but not quite*" (122). Inevitably, "mimicry is at once resemblance and menace" (123).

as an allegory of Lucy's present inability to fit in and her potential "progress" in America. One implication is that she has been too long with the "monkeys" (and so cannot fit into their happy "human" society). While Lizabeth Paravisini-Gebert sees Lewis as an embodiment of "*latent* notions of racial, cultural, and class superiority" (122, italics mine),[26] it seems as if Lewis tries to coax Lucy into a mimicry of ethnoracial[27]/cultural homogeny, as if the only way Lewis (and perhaps Mariah too) can recognize Lucy as human is to see her as a reflection of himself, however "aped" it appears in his mind.

The bad faith of this discourse of animality is manifold, as coded racism *and* untenable speciesism, as the word "actual" implies. The uncle's monkeys aren't just monkeys but ersatz humans. Lewis's cautionary tale suggests they are dangerous, indeed monstrous, because they are so much *like* humans that we may learn to prefer them (such as the monkeys' lovers in *Candide*[28]) or, worse, or lose the ability to differentiate between them and "real" people, turning us into the proverbial monkey's uncle. In this topsy-turvy scenario, that which supposedly makes us human is our ability to differentiate ourselves from our primate relatives, an ability we can unlearn (of course, this ability is not unique to humans—most if not all species of animal differentiate between themselves and other species).

Lucy's response to this story has understandably attracted a number of critics. Jana Evans Braziel reads her dream-narrative as a mediation of her diasporic experience ("Daffodils ..." 122), while Justin D. Edwards sees a misunderstood, "innocent narrative, which mixes together images from *Alice's Adventures in Wonderland* and *The Wizard of Oz*" (70). Closer to the point, Donnell argues that these images are, in fact, of "plantation life and slave capture alongside the yellow brick road," and that the narrative "serves to question the value of dominant psychoanalytic theories, which sees dreams only as ciphers for issues of sexual difference and conflict" denying "issues of cultural difference" (*Routledge Reader* 488–9). In an interview published in *The Missouri Review*, Kincaid's comment on the scene supports this reading: "The people in Lucy's society live for dreaming. They believe that waking life is informed by dream life. ... Louis and Mariah were in fact saying that her perception of the world was not valid, that she needed Freud" (*Conversations ...* 30).

---

[26]   She argues Lewis is "[o]blivious to the racist undertone of any reference to monkeys to people of African descent" (122).

[27]   Philosopher David Theo Goldberg coined "ethnoraces" in *Racist Culture: Philosophy and the Politics of Meaning*, 1993. In *Less Than Human*, which I will discuss in the Conclusion, David Livingstone Smith uses the term because ideas of race are always bound up with ideas of ethnicity, nationality, and more, and because race is about what people "are *thought* to be" (185).

[28]   Though the monkey-lovers scene in *Candide* is not this kind of cautionary tale, despite the "science" and politics of Cacambo's explanation of it: "They are a quarter human, just as I am a quarter Spanish" (73).

Following my reading of Lewis's story as a demand for mimicry, Lucy's dream-narrative may be read as part of a larger "gesture against mimicry" that Donnell notes with respect to concepts of mother and "motherland":[29]

> In the Caribbean the psychology of the colonized has been constructed through the implementation of rigorous hierarchies of shade and class which offer mobility only through linguistic, behavioral and even physical acts of mimicry. Colonial subjects have been termed the 'mimic men' because of the demand placed upon them by colonial institutions—schools, churches and workplaces—to mirror the ways of the colonizer in speech, dress and behavior. ("When Daughters Defy" 22)

After the images of capture and plantation life, the narrative ends with Lucy falling down a hole containing "silver and blue snakes." These snakes suggest a possible allusion to the water snakes of Coleridge's "Rime of the Ancient Mariner" (lines 277–87), interesting in the context of current readings of the poem's skeleton ship as a slave ship and Lucy's relationship to British Romanticism[30]—specifically, to Wordsworth's "I Wandered Lonely as a Cloud," which Lucy (and Kincaid) was made to memorize in school as a symbolic representation of the imperial homeland (and which gave her nightmares).[31] This suggests that while the colonization of culture necessarily extends to the deepest reaches of consciousness, Lucy turns colonial training and knowledge against Lewis's story by responding with her own, a narrative that affirms the history and legacy of slavery and imperialism as the reality beneath his condescending concern, the hole beneath the ground of his subjectivity (a dream that may partially "rewrite" the childhood nightmare instigated by memorizing Wordsworth's poem, that of being chased and buried by daffodils—here Lucy is walking on yellow ground, yellow like the "golden daffodils" and her employers' "yellow heads"). It may also be read as a mockery of Lewis's racist associations, of the association of people of African descent

---

[29] "This life-affirming gesture against mimicry [in *Lucy*] can only be fully appreciated by a cultural reading, in which the power of her actual mother is taken as a metonym for that of the colonial-mother-land and mother-tongue" (22).

[30] In the same interview quoted above, Kincaid said, "A sort of desire for a perfect place, a perfect situation, comes from English romantic poetry. It described a perfection which one longed for, and of course the perfection one longed for was England. I longed for England myself. These things were a big influence, and it was important for me to get rid of them [through writing them]. Then I could actually look at the place I'm from" (Conversations 31). In "Lucy and the Mark of the Colonizer," Moira Ferguson writes, "The fact that William Wordsworth wrote several 'Lucy' poems raises the question of Kincaid's possible ironies. Although we learn … that Lucy is named after Lucifer himself, Wordsworth's Lucy and Kincaid's Lucy are also mirror-like; one mimics the other" (55).

[31] From Chapter Two of *Lucy*: "The night after I had recited the poem, I dreamt, continuously it seemed, that I was being chased down a narrow cobbled street by bunches and bunches of those same daffodils that I had vowed to forget, and when I fell down from exhaustion they all piled on top of me, until I was buried deep underneath them and was never seen again. I had forgotten all this until Mariah mentioned daffodils, and now I told it to her with such an amount of anger it surprised both of us" (18–19).

with monkeys and monkeys with lewdness, as a response to the request implied by Lewis's story that she mimic his "human" (Western) values. In this dream-narrative, Lucy is naked but Lewis is *chasing her* (and with Mariah's approval or at her instigation). In this sense, Lucy's narrative and Mariah and Lewis's response to it (itself a response to Lewis's "monkey" story) call attention to, as Donnell argues with respect to the novel as a whole, "the ways in which certain intellectual spaces remain colonized within Western thought" (*Routledge* 490).

As the novel progresses, it becomes clear that Mariah and Lewis and their friends are, literally and figuratively, the tourists of *A Small Place*. They had "all somehow been to the islands—by that, they meant the place where I was from—and had fun there. … I wished once again that I came from a place where no one wanted to go … it made me ashamed to come from a place where the only thing to be said about it was 'I had fun when I was there'" (65). The connection between the discourse of animality, constructions of race, and globalization as modern day imperialism is realized in the tourist's view of the island as natural playground and the native's happy "harmony with nature."

This static "harmony with nature" is another way of constructing others as "animals," a denial of the existence of whole bodies of knowledge, traditions, and ways of human *and* nonhuman life. This discourse of animality (a "bestiality" that applies only to other species or other human peoples) is both a foundation and effect of the humanist idea of culture. In postcolonial places, the "introduction" of culture to the supposedly cultureless was a result—and a means—of classifying entire peoples, species, and lands as resources for production, commodification, and consumption (again, "we *were* capital, like bales of cotton and sacks of sugar"). As Kincaid asks in *A Small Place*,

> In countries that have no culture or are afraid they may have no culture, there is a minister of Culture. And what is culture, anyway? In some places, it's the way they play drums; in other places, it's the way you behave out in public; and in still other places, it's just the way a person cooks food. And so what is there to preserve about these things? For is it not so that people make them up as they go along, make them up as they need them? (49–50)

The answer here is, as Raymond Williams, Richard Hoggart, and others in cultural studies have contended, culture is ordinary. And, as cultural biology demonstrates, much more ordinary than we think. The "ruling idea" of culture is no different than any other ruling idea, save that it seems the most powerful. One must ask of it what *A Small Place* asks of capitalism and life in postcolonial Antigua generally: "who profits from all this?"[32]

---

[32]   Hoving writes, "Kincaid is much less concerned with the reconciliation [Audre] Lorde strives for, but she is driven by the same question: who profits from all this? This is the question that structures *A Small Place* in different ways" (193).

Dehumanization Takes a Holiday

Virtually everyone who writes about *A Small Place*, whether in scholarly articles or book reviews, remarks on the sharpness of its anger.[33] Ferguson analyzes the "unabounding ire" radiating in the text, concluding that "[i]ncontestable denunciation in *A Small Place* has replaced the implicit jabs of *Annie John*" (*Colonialism* ... 135). The *New Yorker*, for which Kincaid wrote for several years, rejected the text because "it expressed too much bitterness and anger" (Edwards 8), a refusal that speaks volumes given the context of Kincaid's employment with the publication, characterized by Anne Collett as "a trade in the erotic exotic, of which the exchange of 'sassy black' [words] for 'cocktail things' [and, later, income] is signatory" (98). Collett explores the degree to which Kincaid's *New Yorker* writings "disrupt or become party to the globalizing, colonizing 'ethics' of a Northern American cosmopolitanism signified in the glossy consumerism of the *New Yorker*" (99). With Kincaid's acerbic commentary on power and luxury running in the same publication as advertisements for cruises to Caribbean islands, Collett must ask "to what degree does it [Kincaid's column] refuse or embrace what might be seen as the commodification of liberal humanism?" (99). With copy too acidic—meaning, too political—for the *New Yorker*, this is a question one need not ask of *A Small Place*.

And while everyone seems willing to talk about Kincaid's anger, not as many critics discuss the violence of *A Small Place*. The violence of "blow[ing] things up" (32) and the mutilation "of you, your wife, and your children ... in your beautiful and spacious bungalow at the edge of your rubber plantation" (35) gives pause to readers and critics alike. While this violence is *horrific*—the narrator is, after all, talking about the mutilation of children—Kincaid does not approve of but voices and historicizes this violence. What some readers find deeply troubling (including almost all of the students with whom I've read the text over the last fifteen years) is exactly this voice, the voice of the perpetrator of violence speaking to them. Its very proximity is disturbing—it creates anxiety both to hear this voice in one's ear and to hear that this voice is *human* (not monstrous or machine). However, it is exactly this narrative embodiment that, I think, rescues such moments from callous abstraction. It is in this spirit of voicing dehumanization that the narrative dehumanizes the reader/tourist/colonizer as "rubbish" and so on; in the text, the reader ceases to be a person and becomes, instead, an idea ... like "the animal." While *A Small Place* knowingly enacts the dehumanization of the reader/tourist (and, to some degree, native) for political reasons, for the hope of real politics, it is necessarily bound up with the very consciousness it critiques. As Helen Tiffin

---

[33] Examples from scholarly texts quoted earlier in this chapter. As for reviews, Alison Hill called the book "distorted by anger" (Edwards 78). Even Salman Rushdie's glowing review used the words "jeremiad," "force," and "torrential" (Edwards 8).

suggests, "it is the inextricably 'entangled' nature of the cultural products of that [colonial] history which increasingly commands our attention" (58).[34]

In Kincaid's strangely parallel text, *Among Flowers* and, earlier, in parts of *My Garden (Book)*: (hereafter referred to as *My Garden*), the strategic dehumanization in *A Small Place* becomes concretely "entangled" with a liberal, American *freedom* to dehumanize. *Among Flowers* chronicles Kincaid's 2002 journey to Nepal in the company of botanists to collect seeds for her garden in Vermont; *My Garden* includes a narrative of an earlier seed-collecting trip to China in 1998 (complete with the sketch of a monkey on page 215). Much has been written about *My Garden* and the complexities of a postcolonial relationship to gardening and botanical collecting (by critics such as Jana Evans Braziel, Susie O'Brien, Sarah Phillips Casteel, and Rachel Azima), but *very* little about *Among Flowers* (a point I will return to shortly).

The plant acquisition celebrated in *My Garden* leads Jeanne C. Ewert to exclaim that Kincaid is not only "part of modern gardening's customer base" but has now "joined the class of botanical explorers and conquerors in a more direct way" (117). Ewert argues that although

> Not precisely a tourist in China, and yet not there to appreciate or understand local culture, Kincaid's behavior, which she details with remarkable frankness, is very little different from the tourists she criticizes in *A Small Place*. ... Moments in *A Small Place* describing the appalling assumptions of rich American tourists about the lives of the people who inhabit the Caribbean, are repeated in her own travel narrative. ... Enamored of a Chinese landscape she has long imagined, Kincaid frequently recoils from Chinese people. (118)

Indeed, Kincaid writes, "I had come to China to collect seeds, not to be comfortable with what Chinese people did" (192). This attitude is even more freely and frequently expressed in *Among Flowers*, which at the beginning of the narrative casually dismisses any responsibility to understand local people and their customs, as well as other visitors: "One group [of climbers] was from Austria but we decided to call them Germans, because we didn't like them from the look of them ... Germans seem to be the one group of people left that can not be liked just because you feel like it" (28). And at the end: "As American and British people, we felt free to make fun of the Italians but in a kind way. As American and British people we not only made fun of the Germans, we also hated them" (182).

While *A Small Place* repeatedly refers to the tourist as an "ugly" human being (17, 18, 35), ugly because tourism in the Caribbean (and in many other parts of the world) perpetuates the ugliness of the legacy of empire, Kincaid's pursuit of flowers is itself an ugly business. *Among Flowers* seems like a palimpsest. In *A*

---

[34]   This passage is quoted in footnote four of Collett's "A Snake ... ." Collett notes that Tiffin is herself echoing Nicholas Thomas's usage of "entangled" in *Entangled Objects: Exchange, Material Culture, and Colonialism in the Pacific* (Cambridge: Harvard UP, 1991).

Small Place, the ugliness of the tourist, seen through Antiguan eyes, is contrasted with the "unreal" beauty and deprivation of the island seen through the tourist's eyes: "Sometimes the beauty of it seems as if it were stage sets for a play, ... no real flowers could be these shades of red, purple, yellow, orange, blue, white ... no real cows look that poorly as they feed on unreal-looking grass in the unreal-looking pasture, and no real cows look quite that miserable as some unreal-looking white egrets sit on their backs eating insects" (77–8). *Among Flowers* also repeatedly casts the landscape, here the Himalaya, as unreal—as a frame from a child's Viewmaster toy, a calendar, a mirage[35]—but not as a critical reflection of, or on, the tourist's gaze. In this text, Kincaid is more than a postcolonial subject and resident of a wealthy, powerful country; she is also a relatively wealthy tourist. She never, however, refers to herself or her companions as tourists; in fact, Kincaid goes to some trouble to state that they, unlike the outdoor adventurers, have a real purpose for being in Nepal.[36]

During a reading of *Among Flowers* at Brown University in February 2005 Kincaid remarked on "the irony of her 'transgression'; as a woman from the Caribbean she was once looked at 'by the prosperous part of the world',' and now she travels 'from the prosperous part to look at (Nepal)" (Simons); in a public interview with Marina Warner in London later that year, Kincaid qualified this "transgression" when questioned by a member of the audience:

> **Audience** ... You are walking in Nepal but you are in fact in the same kind of 'tourist' position that you critique so harshly in *A Small Place*. ...
>
> **JK** ... the kind of tourists I was writing about [in *A Small Place*] were not, for instance, the tourist who goes to Vienna. ... Well, the reason this seems to me all right is because there is something equal in the relationship, the Viennese can also come here. But the book I wrote about tourism was talking about the continued exploitation of a group of people, but in this other form [tourism]. Now, was I exploiting, was I an eco-tourist, a new concept? Usually the tourist seems to go somewhere and have a good time, generally speaking. I don't think there is much of a good time to be had by a botanist-tourist in this part of the world. I was doing something *slightly* transgressive, but I have to say in the world of things, to condemn this is *a small thing*. So the botanist-tourist, *I'll think of that next time. As I say it's hard, and you could die doing it.* I don't think that any tourist thought that they would die from going to the Bahamas. ("Among Flowers" 56, italics mine)

Kincaid did not think of herself as a tourist in Nepal, despite the fact that she was traveling for pleasure (and to a very poor country, one that is still struggling

---

[35]   Kincaid records at least one similar observation in China, "where the moon looked strange in the sky and the sky itself looked not real" (*My Garden* 215).

[36]   She uses "traveler" to refer to the non-natives she encounters.

120   *Ecocriticism and the Idea of Culture*

for democracy[37]). Her remarks suggest that difficult conditions or the possibility of personal danger mitigate this "slight" transgression. Despite the example of "acceptable" tourism at the beginning of her answer, there is no mention in her response of the overwhelming poverty faced by most people in Nepal (unlike the Viennese or the English, most Nepalese could not afford to visit another country). At the time of Kincaid's visit, the GDP per capita (in current US dollars) for Nepal was $237, compared with Antigua at $10,225, and the US at $36,797.[38]

Among Flowers is, compared to *A Small Place*, a conventional tourist narrative; it is a first-person record of Kincaid's journey, with lists of items taken on her trip, descriptions of souvenirs purchased, and photos of the author with locals, and alone, on a mountain top (see cover). However, it also exposes painful moments of political ambiguity and personal feeling. Kincaid frames her narrative of recreational travel with the recognition of her privileged status: "I experienced many difficulties in this adventure, but they were of a luxurious kind. I did not like, and could not even bring myself to understand, my hosts' relationship to food: I feel that the place in which it is taken in, eaten, be it the kitchen or the dining room, must be far removed from the place, bathroom or outhouse, in which it comes out again in the form of that thing called excrement" (2). This confession echoes the passage in *My Garden* that leads Kincaid to declare that she is not in China to understand the Chinese: "I saw a large family having a wonderful time as they ate their dinner; it was so heartening it made me homesick, and I wanted to join them; but the baby of the family was having a bowel movement on the floor right then; it was all very comfortable for them, but I had come to China to collect seeds, not to be comfortable with what Chinese people did" (192).

Kincaid's admissions of repulsion are almost an invitation to consider critically her inability to understand such difference. She is herself *very* critical (rationally and irrationally) of other travelers in Nepal, from Austrian climbers (noted earlier) to the variety of Westerners who seem to try to blend in with the poverty around them. Kincaid points out that these Europeans, who "look poor, dirty, and bedraggled," are still living a kind of luxury, "for these people are travelers, at any minute they can get up and go home" (17). Kincaid makes a sharp distinction between these people, who consume Nepal for their own pleasure, and herself and her companions, who are in the region to collect seeds for commercial enterprises and their personal gardens—for economic profit and personal pleasure:

---

[37]   Though parliamentary elections were first held in 1959, Nepal is still struggling to form a democratic constitution and uphold basic human rights: "Constituent Assembly convened May 2008 to draft a new constitution. In May 2010, the deadline for the drafting of the constitution was extended 1 year. ... Trafficking in women and child labor remain serious problems ... According to the State Department's 2008 Trafficking in Persons Report, 5,000 to 7,000 girls have been trafficked from rural parts of the country to Kathmandu, and there are over 20,000 child indentured domestic workers in Nepal" (http://www.state.gov/r/pa/ei/bgn/5283.htm).

[38]   According to the World Bank: http://data.worldbank.org/indicator/NY.GDP.PCAP. CD?page=1.

"What they wanted was to collect the seeds of plants that would make a gardener like me, someone who wanted to know about and be engaged with the world but *in the most benign way possible*, excited" (115, italics mine). While Kincaid is collecting seeds for personal use, the ethical distinction between her group and other tourists—mountain climbers, spiritual seekers and temporary residents— is dubious at best. In fact, as Kincaid has demonstrated in other contexts, seed collecting (and gardening those seeds) is not necessarily a "benign" activity.[39] It was the tool of empire (such as the botanical gardens of her youth) and is now part of, as Vandana Shiva suggests, an even more trenchant form of globalized oppression: technological biopiracy.[40] It is difficult to tell from the tone of this passage if Kincaid uses the word "benign" at all ironically. Again, in the context of her oeuvre one might assume so, but this assumption would miss the tension between conflicting and overlapping narrative identities: the wealthy tourist and the benign visitor, the American resident and native Antiguan.

Throughout the book, Kincaid moves between describing food and eating and excretion—as if transplanting the tourist's anxiety about floating shit in *A Small Place*—at times duplicating in the narrative the physical proximity between food and shit she found horrible in Nepal and, earlier, in China.[41] Some of these scenes are enveloped in observation and in her discomfort with being closely observed by local people. At first she seems to accept this as tit-for-tat: "It felt odd but also seemed fair: we were in their country looking at their landscape after all" (38). However, the observation of things that, for Kincaid, should remain hidden, is particularly troubling; it is another kind of uncomfortable proximity: "Not to allow anyone an awareness of the workings of your body is easy to do in our normal lives, where we have access to our own bathrooms ... toilets that allow their contents to disappear so completely that to ask where to could be made to seem a case of mental illness" (91). Though the passage suggests that the mechanized "disappearance" of waste is somewhat unreal, like a kind of magic, and that to question where it has gone sane indeed, before long Kincaid herself feels not only objectified, made unreal, but consumed by the gaze of others: "It seemed as if all the people living in the area had descended on our camp and were just sitting and looking at us. It was as if we were a living cinema. They watched us talk and eat. Sometimes they peered at us ..." (100). Sometimes they watched them excrete (75).

These "unreal" moments are part of the curious disconnect at work in the narrative, exemplified in this scene: "We ate our lunch, fresh vegetables and tinned fish, and some people—inhabitants from Muri or not, we could not know—watched us do so. Some of the children had hair that had lost its natural pigmentation; it

---

[39]   For example, many species transported to new ecosystems have become invasive in their new climate. It is unclear if, among the seeds Kincaid brings home, she transports species that might potentially "colonize" (and harm) her ecosystem.

[40]   See *Biopiracy*, *Stolen Harvest*, and *Earth Democracy*.

[41]   Noted in several passages, including pages 196 and 206–7.

had been black but had become blond, a sign that some essential nutrient was missing from their daily diet" (51). According to the Nepal Ministry of Health, nearly half of children under five in Nepal were malnourished as of 2002.[42] This fact probably isn't lost on this group of botanists and gardeners. And yet, there is no mention of sympathy or aid for the malnourished children who watch them eat. Around the time of Kincaid's travel, rice production began to fall, leading to food shortages that reached crisis levels in 2009.[43] In August of that year Kunda Dixit of *The Nepali Times* wrote, "Sixty percent of children under five in the mountains are undernourished: one of the worst figures in the world."[44]

One might expect this scene to be staged as a dilemma. But Kincaid simply notes that these malnourished children watch her eat and then moves on to describe the rest of her day's walk (much, it seems, as she might have picked up and continued on her actual journey after lunch). This disconnect is, again, part of a larger pattern of disconnection in Nepal, where Kincaid is anxious about the observation of her bodily cycles, about consumption and excretion, even as she repeatedly turns the "banality and boredom" (*A Small Place* 19) of the local struggle for existence into a source of enjoyment for her consumption, from the extreme weather conditions in the Himalayas and the "delicious danger" posed by the guerillas (73), to the difficulties of food production, as in this passage: "This scene of house, barn, outbuildings, did not look prosperous; it looked more like toil and eking out an existence. It looked industrious. ... I enjoyed this scene of familiar domesticity" (52). This domesticity is "uncivilized"—much in the way that Western tourists venturing into local neighborhoods in the Caribbean (depicted, for example, in *Life and Debt*) find it pleasurably "uncivilized." Kincaid writes, "We finally got to Taplejung and it made me sad to think of civilization. It was a crowded maze and a mess" (179).

This pattern of disconnection extends to the people who make Kincaid's tour possible. In the same question and answer session of the interview with Warner in London quoted earlier, Kincaid is faced with an audience member, an employee of Tourism Concern, who asks her what she thinks about "the plight of mountain porters, especially in Nepal, who have come to us and said they feel dehumanized and have had their humanity taken away and become pack animals just so tourists can drink tea on the mountainside, so-called eco-tourists" (56). While Kincaid

---

[42] "Nepal is the poorest country of South Asia. About 42 percent of the Nepali population lived on incomes below the poverty line in 1995–96, 46 percent of the adult population remains illiterate (Central Bureau of Statistics 2003), and almost half the children five years and under are malnourished (Nepal Ministry of Health 2002)" (See the World Bank's "Work-related Migration and Poverty Reduction in Nepal" at http://documents.worldbank.org/curated/en/2007/05/7585341/work-related-migration-poverty-reduction-nepal).

[43] See Reuters, the BBC, and *The Nepali Times*.

[44] "Nepal is now even more unlikely to achieve the UN's goal of eradicating extreme poverty and hunger by 2015." http://www.nepalitimes.com.np/issue/2009/08/07/Nation/16199.

acknowledges that the porters are essential to this form of travel and that they are exploited, she begrudges their right to collective action: "They, in turn, know they are and they *take matters into their own hands.* ... Our porters didn't go on strike, but at a certain point they drank a lot, the bags were late; we had in mind that we would get to our campsite and our clothes and tents would be waiting for us ... but they were often hours behind. Yes, they are exploited, no question and anything *you* can do for them, do" (56, italics mine). The "you" in the context of this response reads like a denial of the tourist's responsibility for her part in systemic inequality.

This "you" echoes Kincaid's denial of the politics of her tour in Nepal in *Among Flowers*. Her claim that her tourist anxiety is personal serves to deny her role in a political system of power. Here, Kincaid, as the first-person narrator, interacts with all but one of the porters as nameless, as nobodies; in other words, as servants (a position that *A Small Place* decries as an extension of slavery, a practice of being a "good nobody"):

> *I could never remember their proper names*, I could only remember the person who carried my bag, and this from looking at his face when I saw him pick up my bag in the morning. This is not a reflection of the *relationship between power and powerless*, the waiter and the diner, or anything that would resemble it. This was only a reflection of *my own anxiety, my own unease, my own sense of ennui, my own personal fragility*. I have never been so uncomfortable, so out of my own skin in my entire life, and yet not once did I wish to leave, not once did I regret being there. (27, italics mine)

While this dynamic between the powerful tourist and less powerful servant may seem only a reflection of "personal" anxiety to Kincaid, it is unlikely that the porters share this perspective. Moments in the narrative like this one seem shocking—not simply because Kincaid makes the commonplace gesture of excusing dehumanizing attitudes or behavior through frank personal admissions or self-conscious statements[45]—but because she does not acknowledge that this tourist's anxiety is never simply personal, that it is a fundamental part of the politics of consuming the Other, and the Other's environment and culture, politics which do not disappear with good will or the admission of unease.

While Kincaid writes that her "luxury" (her tourism) is made possible by porters, and is at times shocked by the "civility" carted over mountains on her behalf, these moments of sympathy feel deeply colonial, as here: "When I saw the man whose job it was to carry the table and chairs wherever we went, I was

---

[45]   For other examples of this move, see her remarks about Germans on pages 28 and 182, quoted earlier, and this statement made in an interview in 1987: "my mother had named me after someone whom I particularly came to loathe—a Lebanese woman, one of these people who come through the West Indies to get something from it but they don't actually inhabit it. I know this sounds awfully racist, but I just can't stand those people" (Cudjoe 218).

appalled that someone had to carry this whole set of civility ... And we could not pronounce or even remember this man's name, and that is how we came to call him 'Table.'" (30). And even more clearly colonial here, when Kincaid wonders how the porters who carry her food come by their own: "They sang songs and made and ate their food along with us. I realized then, that I had no idea how or when they acquired or ate their food. They sat around with Sue and me and she taught them how to separate various seeds from their fleshy fruit. *They knew nothing*" (152, italics mine). Kincaid records trouble with these "undisciplined porters" (49), who become drunk (158) or obstinate (162), throughout the book. Without Kincaid's interview comments, one might be tempted to read this narrative as a political appropriation of and reflection on the neocolonial attitude and voice of the tourist (as we see in *A Small Place*). Indeed, as we will see in a moment, Murray argues just this. But, as Kincaid herself suggests, she does not think of herself as a tourist in Nepal, despite moments when she is alive to the fact that these porters serve her pleasure, as here: "What had the porters been doing all day? someone said—meaning, What had they been doing when we were exploring the landscape, looking for things that would grow in our garden, things that would give us pleasure ... and remember seeing them alive in their place of origin, a mountainside, a small village, a not easily accessible place in the large (still) world?" (83).

Kincaid's sensitivity to the insensitivity inherent in her position manifests itself as bad faith near the end of her journey and narrative when the porters' "sudden rebellion" leads her to imagine their thoughts about herself and her companions: "we wondered if all along what we had thought were encouraging words, spoken to us in their native language, was really them mocking us, finding us and our obsession of their native plants ridiculous, worthy of jokes made just before they fell asleep" (162–3). This echoes, rather uncomfortably, the native's view of the tourist in *A Small Place*: "the people who inhabit the place in which you have just paused cannot stand you ... behind closed doors they laugh at your strangeness"— even worse, these locals may "collapse helpless from laughter, mimicking the way they imagine you must look as you carry out some everyday bodily function. They do not like you. *They do not like me!* That thought never actually occurs to you" (17). *Among Flowers* at times gives the incredible impression that just having this thought (and publishing a photo of a porter carrying a heavy load, on page 82) may itself be enough or, worse, may be all that is possible—that the recognition of the existence of other forms of consciousness and culture are the only "politics" left to those of us who *enjoy* such privilege.

The political anger of *A Small Place* is, sadly, nowhere to be found. The only anger in *Among Flowers* is "personal"—in response to Kincaid's treatment at the hands of the porters and the Maoists. Kincaid's references to, and depictions of, Maoist guerillas seem not only unsympathetic and unhistoricized but, at times, dehumanizing, as here, "At some point I stopped making a distinction between the Maoists and the leeches, at some point they became indistinguishable to me, *but this was only to me.* Fortunately I had acquired some DEET" (90, italics mine). Kincaid's depoliticization of the conflation of the "army of leeches" and Maoist rebels ("but this was only to me") echoes or enacts the lack of political consciousness about racism and dehumanization she exposes in *A Small Place* ("Our perception of this Antigua … was not a political perception"). In the end, Kincaid "remembered the leeches more than … the Maosists …" (86). She asks herself, "What was I doing in a world in which king and Maoists were in mortal conflict?" (20) and answers—repeatedly—pursuing *pleasure*. She does, however, briefly consider the Maoists and the porters together:

> The Maoists were right, I felt in particular: life itself was perfectly fair, people had created many injustices; it was the created injustices that led me to being here, dependent on Sherpas, for without this original injustice, I would not be in Nepal and the Sherpas would be doing something not related to me. And then again, the Maoists were wrong, the porters should be fired; they were not being good porters. They should bend to our demands, among which was to make us comfortable when we wanted to be comfortable. … We wished Sunam would fire the porters. But he couldn't even if he wanted to. There were no other porters around. (84)

This passage presents, in miniature, the overall problem with *Among Flowers*. While Kincaid is clearly critical of her "demand" for comfort, she still demands comfort (and the narrative almost seems to present this as just another fact of life in an inevitably unjust world because Kincaid is, at times, sensible to injustice).

While Kincaid's strategic dehumanization of the tourist in *A Small Place* becomes entangled with the liberal "freedom" to dehumanize on holiday, displayed in *Among Flowers* (and *My Garden*), dehumanization is itself entangled with a discourse of animality and practice of speciesism. As she writes self-consciously in the latter, "we saw some white-haired monkeys way above us in the trees, and they made the most wonderful sounds to each other. I was so happy to see them; and this suspicious thought crossed my mind, that I was happy to see them because to see them is to claim them. *Claiming, after all, was the overriding aim of my journey*" (71, italics mine).[46] One way or another, nonhuman animals, porters, villagers, landscape, and seeds have all been "claimed" on this journey. As the audience member from Tourism Concern told Kincaid, "they [the porters] feel dehumanized and have had their humanity taken away and become pack animals just so tourists can drink tea on the mountainside, so-called eco-tourists."

Very few critics have written about *Among Flowers* (though it was published in 2005), which is remarkable given the amount of criticism on Kincaid's other works, including her other "garden" writing. Murray's *Island Paradise* includes

---

[46] Kincaid records this event a few pages later: "All the children seemed to have walked up to a ledge that was right above us, and they climbed into the trees and began to make the sounds that some monkeys, who were also above us in the trees, were making. It was meant to disturb us but it didn't at all. Nothing could be more disturbing than sleeping in a village under the control of people [Maoists] who may or may not let you live" (75).

only four brief pages on the text; these emphasize Kincaid's care "not to objectify" locals and her "discomfort at using porters" (92). She sees *Among Flowers* as a successful reversal of "the traveler's gaze by situating herself and her fellow travelers as the object of inspection," part of an ongoing "negotiation between the powerful and the powerless" (94). It is not surprising that Murray does not mention any of the many truly troubling passages in the text. One must skip over a great deal to read *Among Flowers* as a challenge to the imperial gaze.[47]

For more substantive critiques of Kincaid's botanic-travel writing, one must look to work on *My Garden*. Isabel Hoving argues that Kincaid's gardening practice eschews representation and avoids "an anti-colonial discourse that would accuse and demand amends. In her account, there is no use for binary ideologies that would oppose a violent, alienating, colonial practice of up-rooting and transplanting to a peaceful, native practice of natural gardening" (132). And this makes sense with respect to gardening; however situated a practice, it involves some uprooting and transplanting—indeed some killing. It can be, to varying degrees, a practice of mastery.[48] And both Hoving and Murray (and others) discuss the way in which an idea of an original garden, of Eden, "became a justification for land-theft and colonization" (132). Hoving argues that instead of a colonial or anti-colonial aesthetic, Kincaid resorts to a "sensual discourse ... of pleasure ... in the complexity of lived space" (132), practicing an "anti-colonial anti-aesthetics" (134). This practice is "inevitably politically aware, and aiming at political insights; but also eager for the idiosyncratic" (135).

Hoving contends that Kincaid does not oppose home and travel as "objects or conditions" but engages them as "functions, adjectives, verbs" (136). She reads Kincaid's ethics, her view of all knowledge as situated and problematic, her resistance to neat oppositions, as a tarrying with the universal: "*linger*[ing] in the opacity of colonialism ... is an ambiguous desire ... more radical than a consistent critique of Enlightenment, and more soberingly practical" (138). In short, she suggests, Kincaid presents us with idiosyncrasy, or an invitation to "approach the universal as a 'non-generalized universal'" (Melas qtd. in Hoving 138). Though Hoving does not address the narrative of Kincaid's visit to China published in *My Garden*, she praises the "complexity, thickness, and opaqueness" of her work,

---

[47]   Amazingly, Ashmita Khasnabish argues that the text is a wholly benevolent engagement with people and the world: "*Among Flowers* truly reflects the genuine note of globalization. She inhabits a global space indeed; in her world in the Himalayas each and every person seems to be very close to her. She does not differentiate between them and herself in terms of nationalities or geographical boundaries. She seems close to Sherpas, Maoists, and the pedestrians she meets" (para. 13). She concludes, "Kincaid's *Among Flowers* points toward a humanitarian identity, in which the world can come together through various paradigms. She achieves it through nature, for which the feminists came up with the term eco-feminism, and the whole book of her journey through the Himalayas confirms that" (para. 17).

[48]   As I argue in "Biogenetic Intervention (Or 'gardening,' Shakespeare, and the future of ecological thought)." *Green Letters* 9 (2008): 33–47.

which refrains from (and, presumably, warns against) "any quick understanding" and "against the imagination and existing textual discourses as a means to understanding the postcolonial" (135). The Hoving hones in on Kincaid's practice of home (rather than her practice of tourism), her transnational interconnectedness in place, and suggests how we ought to read this practice: "the academic quest for knowledge should respond though intimate interaction to messy embodiments like this active home [depicted in *My Garden*]" (138).

It is more difficult, I think, to respond as cleanly to Kincaid's "messier" embodiment in China or Nepal. Kincaid's pleasure in Nepal does not cohere into the "anti-colonial anti-aesthetics" of her (home) garden (which, since it grows from these trips, becomes suspect as well). While sobering, these botanical journeys aren't "soberingly practical." In fact, practical politics is exactly the problem. Kincaid's acknowledgement of her privilege is critique *sans critique*. Scott writes,

> *My Garden Book* enunciates from a position of privilege. ... the voice is that of the tourist looking at the natives—poor peasants in a huge country characterized by extreme uneven development—with scorn and disgust worthy of a Waugh in the Caribbean. The most striking theme of this section is horror at the pervasive filth and excrement. Rather than seeing these as symptoms of poverty, the narrative discusses them in cultural terms as evidence of different standards ... (75)

Scott reads these parts of texts as, at best, the product of a liberal (soft) multiculturalism rather than a materialist understanding of socioeconomy; so, while *A Small Place* depicts class as a relationship, *My Garden* (and, I would add, *Among Flowers*) posits "class ... as identity: [in which Kincaid's own] moving from a position of want to one of comfort is conceptualized as joining the 'conquering class'" (77). In this way, Kincaid's pursuit of flowers remains disconnected from the need for "bread and roses" despite the historical power relations involved in botany.

This analysis of *Among Flowers* should not, however, be read as a challenge to Kincaid's motives, as Ewert does here with respect to *My Garden*: "While the evidence for colonial abuse and enslavement of both plants and native gardeners that Kincaid marshals is incontestable, when she describes her own garden the reader is forced to question the motives behind her outrage" (116–17). The problem seems less a question of intention than action; it is the materiality of Kincaid's tourism in China and Nepal (as she suggests in the interview with Warner with respect to Europeans or Americans in Antigua) that entangles her earlier travel writing. Discussing Kincaid's incarnation of the tourist of *A Small Place* in China, Ewert quotes Barbara Edlmair's argument that, even in this earlier text, "the ambiguity of her [narrative and material] position is never problematized"[49] (124), concluding that her "hybrid garden writing leaves perhaps less room for political optimism" (125).

---

[49] Eldmair 79.

The narrative embodiment that rescues the violence of *A Small Place* from callous abstraction (the voice of the perpetrator of violence speaking to us as a human voice, in history) is entangled with self-conscious but depoliticizing embodiments of the tourist/colonizer in *My Garden* and *Among Flowers*, from Kincaid's own anxiety of embodiment to her "animalizing" view of other native peoples, nonhuman animals, cultures, and ecosystems. While the dehumanized reader/tourist/colonizer in *A Small Place* becomes through strategic reversal an idea, like "the animal," the discourse of animality (which is always also one of human embodiment) remains largely unexamined within Western and postcolonial theatres of exploitation, anxiety, and containment.

Speciesism, grounded in the fear of human animality and nonhuman beings (the complex diversity of agential subjects and density of material politics), is played out in an abyssal mimicry and the demand for mimicry, where colonizers and the colonized are configured as predators and prey, disembodied subjects and liminal primates. As I will argue in the Conclusion, the critique of the discourse of animality does not elide or discount the disempowerment of human beings but the reverse; meaningful recognition of the continuity and deep interconnectedness of species demands political continuity within and beyond the human community. To challenge dehumanization, we must not say "we aren't animals" but, instead, "we are all animals" or, as Derrida would have it, "the animal does not exist," confronting both far-right ideologies and liberal humanism where they live—in the idea of the human.

In "Remembering the Limits: Difference, Identity and Practice," Gayatri Spivak characterizes the justification for colonization as a humanist discourse of development: "That's why all of these projects, the justification of slavery, as well as the justification of Christianization, seemed to be alright; because, after all, these people had not *graduated into humanhood*, as it were" (229, italics mine).[50] It is perhaps for this reason that the genre of the Bildungsroman has been so important to postcolonial writers. According to Antonia MacDonald-Smythe,

> The restlessness and dissatisfaction that initiate the literary subject's movement away from the familial and the familiar and the anxiety to become an articulate and articulating self are widespread anxieties. ... the *Bildungsroman* has come to be identified with a western model of development, which, in inscribing the white male as the universal protagonist, theorized his particular progress into adulthood as prototypical. ... Reviewed, recast, and rendered more applicable to the discursive formulation of the marginal subject, the *Bildungsroman* in the

---

[50] In *Animal Rites*, Wolfe responds that it's "understandable, of course, that traditionally marginalized peoples would be skeptical about calls by academic intellectuals to surrender the humanist model of subjectivity, with all its privileges, at just the historical moment when they are poised to 'graduate' into it. ... [However, if this model remains intact] the humanist discourse of species will always be available for use by some humans against other humans, as well, to countenance violence against the social other of *whatever* species" (7).

twentieth century contextualized the journey of the Caribbean protagonist into adulthood, framing a development of voice and agency within an experience of conquest and domination. (29)

The postcolonial interest in the Bildungsroman is in part a response to this very humanist discourse of development (epitomized, however challenged, by this genre); the dehumanized "regain" their humanity not only by voicing but also by politicizing their dehumanization. However, this anxiety to articulate and to demonstrate articulacy, is in both Western and postcolonial contexts also a deep anxiety about human animality and the nonhuman. Sabine Broeck is right when she writes that Kincaid's writing "interrupts white people's preying on the island [of Antigua]" (822), and that her work "could be read as a devastation [of] as much as a generative effort to put whiteness—as a crucially missing signifier— into Adorno's project of enlightenment critique" (841). But while Kincaid's work critiques Enlightenment thought and ideas, it is also entangled with its most foundational error—not a missing signifier but the missing subjects of signification, the other signifying subjects of nature.

Chapter 6
Conclusion:
Dehumanization, Animality,
and the Bildungsroman

In March of 2011 *Der Spiegel* published photos of United States Army soldiers posing with corpses in Afghanistan, corpses positioned as "trophy" kills; in April of 2012 *The Los Angeles Times* published more such pictures of US soldiers, this time posing with the severed hands and legs of Taliban members.[1] These and other similar photographs, such as the infamous pictures of guards posing over corpses and prisoners positioned as domesticated animals at Abu Ghraib, appear during a time of growing popular, scientific, and critical interest in violence and dehumanization[2] and its seeming opposite, empathy—from the discovery of mirror neurons to the recognition of cross-species empathy. Both the philosophy of dehumanization and the mainstream science of empathy participate in an unexamined discourse of animality. I will argue that our ideas about empathy are bound up with this discourse and vice-versa, a connection that reveals a cultural fantasy of detachment, a story key to the fantasy of the human itself. Dehumanized people are perceived as nonhuman in a very particular way: as agents that lack

---

[1]   In March of 2011, news media around the world followed suit, from *The Guardian* to *Rolling Stone*. *The Guardian* published one of these photos in a story under the headline, "Photos show US soldiers in Afghanistan Posing with Dead Civilians: 'Trophy' pictures show US soldiers posing with corpses of Afghan civilians they are accused of killing for sport"—these soldiers also kept body parts as trophies.

[2]   For example, the popular and academic interest in zombies seems at an all-time high. The paperback *U.S. Army Zombie Combat Skills*, of which the "Dept. of the Army" is a corporate author (the top three listed items by this author are: *U.S. Army Zombie Combat Skills*, *Russian Combat Methods in World War II*, and the *United States Army Survival Manual*) suggest that these members of the 5th Stryker Brigade (the soldiers in the 2011 photos) and the 82nd Airborne's Fourth Brigade (the soldiers in the 2012 photos) were primed for their actions before combat; the U.S. military is actively using the popular phenomenon of zombies to market itself, and with some success. For an academic example from the sciences, see an "epidemiology" of zombies, "When Zombies Attack!: Mathematical Modelling of an Outbreak of Zombie Infection" by P. Munz, I. Hudea, J. Imad and R.J. Smith in *Infectious Disease Modelling Research Progress. And for just one example from the humanities, see John Lutz's "Zombies of the World Unite: Class Struggle and Alienation in Land of the Dead" in The Philosophy of Horror. On empathy or the lack of it, there has been a great deal of work recently, including Simon Baron-Cohen's Zero Degrees of Empathy,* Marco Iacoboni's *Mirroring People,* Martha Stout's *The Sociopath Next Door,* and Jon Ronson's *The Psychopath Test.*

empathy, that is, the ability to sympathize. Anti-Semites claim Jews are "parasites" who would "sell their own mothers" given the chance; racists of several varieties routinely state that members of group X "don't feel for each other like we do"; nature is "red in tooth and claw." This story about empathy is foundational for the Western discourse of animality and the humanist idea of culture.

This conclusion will begin with an examination of the connection between empathy and animalization, circling back to the question of "the animal" and the Bildungsroman's formative role as a genre of formation: the humanist origin story of culture. This in turn will allow a brief return to Kincaid's *My Garden* and a consideration of "gardening" in the previous texts, as well as some recent remarks about ecology and the biological idea of culture.

Empathy and Animality

Philosopher David Livingston Smith's *Less Than Human* is inspired by a question he does not answer: "being [perceived as] human can't be the same as looking human" which "invites us to ponder the question of what exactly it is that dehumanized people are supposed to lack" (28). Because dehumanization isn't unique to one culture, set of cultures, or modernity, Smith does not treat this as a cultural but a biological question. However, Smith does not simply claim a biological basis for dehumanization, he constructs a biological determinism: he argues humans have an inherent tendency to essentialize others, leading to pseudospeciation—the positioning of humans from other cultures as other species. We are, he argues, "*natural-born* essentializers."[3]

---

[3] "There is a substantial body of research," he writes, "showing that human beings are *natural-born* essentializers. We spontaneously divide the world into natural kinds to which we attribute hidden essences" (100). This essentialist explanation of essentialism is soon followed by the denial of meaningful continuity between the species. While these ideas might, at first, seem strange bedfellows (biological determinism and human exceptionalism), both reflect a mistaken idea of evolutionary thought. This is where Smith's readers follow him down the rabbit-hole of Western culture, evidenced in his disciplinary predisposition, in which recognizing the mental states of other animals equals conversing with white rabbits in suits and smoking caterpillars. This story, like most origin stories, relies on the binary of nature and culture, the very "yawning dichotomy" Smith occasionally critiques. Despite claiming to know better, Smith takes a good deal of trouble to make it clear to us that human beings are truly separate from the rest of nature, that other animals do not have culture. While some, such as chimps, have "traditions [that] are probably continuous with (although obviously far more rudimentary than) human culture" (66), "there's no reason to think the vast majority of nonhuman animals have the notion of 'us' and 'them': they *react* to organisms as 'other' without *conceiving of them* as such" (125, italics in original). He notes that chimpanzees "appear be an exception" (125), but then argues that the phenomenon of "dechimpization" (also known as chimpanzee war, chimpanzee ethnic cleansing, and so on), involves "little forethought" (210) and so the violence associated with dechimpization cannot be compared to human war, etc. Any concepts chimps may

The "question of what exactly it is that dehumanized people are supposed to lack" is, in fact, the question of how human cultures define humanity. In other words, what makes us other than "the animal"? Dehumanized subjects are perceived as nonhuman animals[4] because nonhuman animals are the ever-ready-made example of sub-beings to whom morality does not apply precisely because "animals" are supposed to lack "fellow feeling" (and so, paradoxically, this is used to justify our lack of feeling for them, our exploitation of them). In other words, the "something" dehumanized people supposedly lack isn't necessarily, or only, what has traditionally (philosophically, scientifically, commercially, gastronomically) been used to separate humans from other animals—language or reason or a dualistic theory of mind—but, I argue, general feeling for others, the source of so-called "humane" behavior, of "humanity" itself.

This claim about other animals is, one might argue, the paragon of bad faith; like Kincaid's tourist, we know (human cultures have always known) this is not true. There are countless observed examples of nonhuman animal empathy, acts

---

have are somehow not really concepts at all—they're fixed and "biological"—"in the chimpanzee mind the division between the two [us/them] is rigid, static, and biologically driven" (213). Chimps—and other animals—are on the other side of what indeed appears to be a yawning dichotomy, an "unbridgeable gulf. ... We are, to a very significant degree, architects of our own destiny. This isn't to say that human malleability is limitless ... *But the powerful intellect with which natural selection endowed us enabled our ancestors to make the momentous discovery that they could engineer their own behavior.* The device they invented for this purpose is what we call *culture*" (125, italics in original). The long italics say it all: the voice of culture articulates with deliberate clarity, "We are not really animals anymore—not like them. *We've invented culture; we make ourselves. Their lives our determined; ours are free.*" And this is the problem with Smith's omission of a critique of a *discourse of animality*, despite his recognition of the West's special preoccupation with the idea of primates, from the writings of Albert Magnus to the "display" of Ota Benga at the Bronx Zoo, and his categorization of the many nonhuman animals used to create a rhetoric of dehumanization: "They are lesser biological entities—typically, dangerous predators, poisonous animals, carriers of disease or disease organisms, parasites, traditionally filthy animals, or bodily products, especially feces" (158). Smith does not account for the way in which all these *things* go together as *ideas*, but a theory of the discourse of animality, as the creation and perpetuation of an idea of the human, does. In this story, nonhuman animals are also the human body, the biological—the opposite of culture, the unculturable.

4   Almost always as nonhuman animals but, very rarely, as plants perceived as agents; to me, such plants are quasi-animal, suggesting what we consider to be "animal" aspects of materiality. While Smith remarks that targeted people are never thought of as "charming animals like butterflies and kittens" but as "animals that motivate violence" (223), even kittens are still "beasts," associated in many parts of the world with "vermin" and in the West with coyness, with an "invitation" to sexual violence against women (think of "sex-kitten" and the metonymic "pussy"), a form of dehumanization Smith does not want to discuss. It is this very idea of "the animal" that "motivates" (legitimizes) violence. As Kincaid said, "when people say you're charming you are in deep trouble."

of sympathy both within and between species.[5] De Waal's *The Age of Empathy* demonstrates that empathy is an evolved capacity that we share with *many* animals, from primates to rodents. He explains the relationship between empathy and sympathy, often conflated, as follows: while empathy "requires first of all emotional engagement," it "is the process by which we gather information about someone else. Sympathy, by contrast, reflects concern about the other and a desire to improve the other's situation" (72, 88).[6] As a number of primate cultures, including human, demonstrate, empathy and violence are not mutually exclusive:

> The violent nature of chimps is sometimes used as an argument against their having any empathy at all. ... There exists in fact no obligatory connection between empathy and kindness, and no animal can afford treating everyone nicely all the time: Every animal faces competition over food, mates, and territory. A society based on empathy is no more free of conflict than a marriage based on love. (44–5)

De Waal discusses evidence of sympathetic behavior or *"targeted helping*, which is assistance geared toward another's specific situation or need" in primates and many other animals, such as seals, elephants, and dolphins (92).[7] He concludes that, "Only the most advanced forms of knowing what others know *may* be limited to our own species" (100, italics mine).[8] "Sharp" lines are "in fact ... sand castles that lose much of their structure when the sea of knowledge washes over them" (149).

Whether we are talking about self-awareness, perspective-taking, empathy, reciprocity, or targeted helping, de Waal cites a variety of evidence that establishes a meaningful continuity of sentience and social feeling between humans and other animals.[9] Those biologists who maintain, for example, that human cooperation is "a

---

[5]   For several examples, see Bekoff's *The Emotional Lives of Animals.*

[6]   Contrary to what we might think, "imagination is not what drives empathy. Imagining another's situation can be a cold affair, not unlike the way we understand how an airplane flies. Empathy requires first of all emotional engagement" (72).

[7]   Targeted helping requires perspective taking, which makes cross-species examples of this behavior particularly interesting. "I find help that crosses species barriers most intriguing, including cases of apes saving birds, or of a seal rescuing a dog. The latter happened in public view in a river in Middlesborough, England" (*Age* 129).

[8]   "I seriously doubt that we, or any other animal, can grasp someone else's mental state at a theoretical level. ... taking someone else's perspective is not limited to human adults. It is best developed in animals with large brains, but those with smaller brains don't necessarily lack the capacity" (*Age* 98).

[9]   For just a few examples, perspective-taking in baboons (Barbara Smuts on baboon vocal consolation, 150), self-awareness in Corvids (149–50), chimpanzee reciprocity and widespread cooperation among nonrelatives (173–80), and pointing ("referential signaling") in other animals, especially other primates (151–4). On pointing: "Inevitably, academics have surrounded pointing with heavy artillery. ... The first step is to move away from silly Western definitions, such as the one requiring the outstretched index finger. In our own species, too, a lot of pointing is done without the hands. ... Monkeys, too, often point with their whole bodies and heads when they recruit allies during fights" (151–2).

'huge anomaly' in the natural world," de Waal writes, are not "anti-evolutionary—on the contrary, they are self-proclaimed Darwinists—but they are eager to keep the hairy creatures on the sidelines. I have only half-jokingly called their approach 'evolution sans animal'" (179). Here, de Waal echoes aspects of Haraway's critique of Western primatology as "Simian orientalism": the Western "construction of the self from the raw material of the other, the appropriation of nature in the production of culture, the ripening of the human from the soil of the animal" (10–11). While Haraway's observation that de Waal naturalizes capitalism in *Chimpanzee Politics* holds, in his later works[10] de Waal demonstrates that unlike some branches of mainstream primatology, cultural biology does not traffic in what Derrida terms "the animal" (as he has it, more to follow).

While he writes about empathy as it connects to both violence and the distribution of resources among human and nonhuman animals, de Waal, like Smith, does not identify a discourse of animality connecting and undergirding both the poor treatment of human animals (poverty, war) and the poor treatment of nonhuman animals. So, while de Waal challenges anthropodenial here and elsewhere, his imperative "to expand the range of fellow feeling" (*Age* 203) is made largely on the basis of human self-interest: "the firmest support for the common good comes from enlightened self-interest … Since empathy binds individuals together and gives each a stake in the welfare of others, it bridges the world of direct 'what's in it for me' benefits and collective benefits" (*Age* 223). Even if this were fine for humans (empathy does not always do such a good job of binding us together, as de Waal himself demonstrates),[11] what about animal-others, the other animals? What if our "interests" conflict with theirs? It is not just a matter of redefining those interests to be more "enlightened"—sometimes our legitimate "interests," our preservation, will conflict with the preservation of other life forms.[12] The humanist discourse of interests, based in liberal economic theory,

---

[10] For example, see de Waal's research on "inequity aversion" or, in simpler terms, a preference for a fair distribution of resources. De Waal's own food experiments with monkeys demonstrate that inequity bothers them; he claims while studies on "inequity aversion in animals have only just begun. … I expect it in all social animals" (*Age* 192). While de Waal dangerously skirts the error of "naturalizing" capitalism ("We look at them [monkeys] as little capitalists with prehensile tails" [195]) or, as Haraway writes of *Chimpanzee Politics*, reading primates as "model yuppies" (*Primate Visions* 128), *The Age of Empathy* does make the important point that economic inequality in the United States has reached dangerous proportions, a process and product of a culture that does not foster empathy.

[11] Both Smith and de Waal note soldiers' "natural" aversion to killing; this, however, leads them to turn their "empathy switch completely down to zero" (*Age* 218), something also achieved through computer-game-like war technology (because empathy is rooted in the body, proximity is important) (*Age* 219–21).

[12] By this I do not mean the kind of question Tim Luke poses in "The Dreams of Deep Ecology" (qtd. by Wolfe in *Animal Rites*, 26) "Will we allow anthrax or cholera microbes to attain self-realization in wiping out sheep herds or human kindergartens?" (51). I mean

does not, much as Cary Wolfe demonstrates with respect to the discourse of rights, admit a solution to the problem of human-caused animal (nonhuman and human) suffering.[13] Political relations, the hope of sharing together the polis we live in, requires more than discourses of interest and rights; it requires a challenge to the idea of the human itself, the discourse of animality that fuels the animalization of human and nonhuman animals.

This discourse relies on a story of detachment, a fantasy[14] that *depersonalizes*[15] human and nonhuman beings. It is a self-serving story about detachment, both ours (as freedom, more on this to follow) and theirs (as a lack of feeling). In the West, it is in fact the very definitions of the *human* (free will, freedom to surpass nature, freedom to master) and the *beast* (unrestrained drives, instinct without empathy): *they (other animals, human Others) do not feel things like we do; they do not have fellow feeling, therefore we are not required to extend to them our fellow feeling.* As de Waal writes, "Warfare and aggression are widely recognized as biological traits, and no one thinks twice about pointing at ants or chimps for parallels. It's only with regard to noble characteristics that continuity is an issue, and empathy is a case in point" (*Age* 207).[16] It is, of course, we who fall short on fellow feeling, of all kinds.

---

more immediate conflicts of interest, daily conflicts between us and our evolutionary "neighbors." Americans, for example, kill countless numbers of cows, pigs, and other mammals because people think (or like to think) their health requires it; it does not, but what if it did?

[13]   See *Animal Rites*. As W.J.T. Mitchell summarizes in his forward, Wolfe demonstrates that "'the rights of animals' ... can be modeled only on the 'rights of man' ... [but] the rights of man are in turn based in the lack of rights in the animal" (xiii). De Waal gives the abolition of slavery in America as an example of progress through enlightened self-interest, specifically of how empathy attached "emotional value" to "collective benefit": "One of the most potent weapons of the abolitionist movement were drawings of slave ships and their human cargo, which were disseminated to generate empathy and moral outrage" (*Age* 224–5). However, again, the subject of slavery provides little evidence for human progress, as it continues worldwide—"even" in America, albeit illegally; Benjamin Skinner's *A Crime So Monstrous* estimates that, as of 2008, there were 27 million slaves worldwide, including the United States. And, in many cases, the descendants of freed slaves are, as Kincaid writes in *A Small Place*, free only "in a kind of way." As de Waal himself acknowledges, "Social Darwinism may be dismissed as old hat, a leftover of the Victorian era, but it's still very much with us" (*Age* 204).

[14]   Similarly, Derrida argues that the idea of "the animal" is a fable: "every animal, as distinct from *l'animot* [animals in all their difference, multiplicity, complexity, etc.], is essentially fantastic, phantasmatic, fabulous, of a fable that speaks to us and speaks to us of ourselves, especially where a fabulous animal, that is to say, a speaking animal, speaks of itself to say 'I,' and in saying 'I,' always, *de te fabula narratur* ..." (66).

[15]   Following Édouard Glissant's use of the term in *Le discourse antillais*.

[16]   Smith does the reverse here; he claims chimps and ants do not make war, that concepts and cruelty are uniquely human (for example, see pages 208 and 214). This claim is, however, based on speciesist logic.

Smith discusses a series of passages from David Hume's *An Enquiry Concerning the Principles of Morals* (1751) that illustrates my point. Like de Waal, Hume argues that morality is based on feeling, and uses the term sympathy to refer "to an inborn tendency to resonate with others' feelings—to suffer from their sorrows and to be uplifted by their joys" (Smith 50). As Darwin, Freud, and others suggest, this *fellow feeling* is the glue—one might even claim the *substance*—of society.[17] It is the basis for the idea of justice: the sense of the treatment due those with fellow feeling (those who feel as we do). Hume argues that, "Because nonhuman animals cannot participate in society, the notion of justice is inapplicable to them" (qtd. in Smith 53). Imagination it seems, particularly our ability to attribute mental states to others, is key; for Hume as for Percy Shelley after him,[18] there exists an "intimate connection between sympathy and imagination. Hume held that it is only because we can imagine that others are beings like us that we can 'enter ... into the opinions and affections of others, whenever we discover them'" (Smith 53). However for Hume, anthropomorphism (including, as de Waal has it, "animalcentric anthropomorphism," the identification of real similarities between human and nonhuman animals[19]) is an imaginative error (Smith 54).

We avoid imagining, avoid knowing, that other animals have fellow feeling to avoid extending ours to them. We do not allow them to participate in human society as *persons,* though they participate in all sorts of other ways—as meat, as slave labor, as discourse and ideology (and this seems far from unrelated to Hume's oft-quoted footnote on the superiority of white people).[20] Despite or, perhaps, because of Smith's training as a philosopher, his analysis does not ask *why* we think of animals as sub-beings; to him it is self-evident, despite his reminders not to think of evolution as teleology.[21]

Psychiatrist and neuroscientist Simon Baron-Cohen's *Zero Degrees of Empathy* (2011) provides an example of work on the subject from the sciences. Baron-Cohen begins his book on the neuroscience of what he calls "the empathy bell curve" with an anecdote about his shock at hearing from a professor of physiology that "the best data available on human adaptation to extreme cold had been collected by Nazi scientists performing 'immersion experiments' on Jews and other inmates of Dachau Concentration Camp" (1). In his own work, however, Baron-Cohen makes use of neurological and behavioral data gathered

---

[17] See *The Descent of Man* and *Civilization and its Discontents.*

[18] See Shelley's *Essay on Love* and the discussion of "Mont Blanc" in Chapter 3. As noted earlier, de Waal takes a different view. For him, empathy is, first of all, bodily (see his discussion of "emotional contagion"); imagination does not drive empathy.

[19] Discussed in the Introduction; see *The Ape and the Sushi Master* (77).

[20] "I am apt to suspect the negroes and in general all species of men ... to be naturally inferior to whites" (qtd. in Eze 33).

[21] "We still unblushingly speak of organism(s) being higher or lower on an evolutionary scale, and the assumption that our species is more highly evolved than others continues, after all these centuries, to suffuse the zeitgeist" (41).

at the expense of the extreme suffering of nonhuman primates (such as the work of the infamous ethologist Harry Harlow[22] and work by neuroscientist Giacomo Rizzolatti). Baron-Cohen's discussion of Harlow's "seminal" work contains only the briefest acknowledgement that this research is "ethically questionable" (49), relegated to a footnote:

> As an aside, it is interesting as to who judges an experiment as unethical. In chapter 1 I was clearly condemning of the Nazi experiments that tested how long a person could tolerate freezing water, yet here I seem to be willing to justify Harlow's and Hinde's monkey experiments. I suspect I am guilty of a double-standard when it comes to human vs. animal research, and that some would adopt an even more stringent view on the ethics of animal experimentation. (147)

This footnote is the only discussion of the animal suffering that makes Baron-Cohen's work possible,[23] one that does not include the words "suffering" or "cruelty." It is important to acknowledge that, paradoxically, Harlow's torture of scores of infants revealed that monkeys, like humans, have and need empathy; all primates (and most likely all mammals and perhaps other animals) need to give and receive affection to become properly socialized. I will conclude this section on empathy where de Waal's book on the subject begins, with his refutation of several Western origin stories. He debunks myths about human beings, such as "our ancestors ruled the savanna" when, in fact, we "lived in terror" of other animals (18) and "human society is the voluntary creation of autonomous men ... they decided to give up a few liberties in return for community life" (20), when in fact society isn't a supplement but a foundational part of our consciousness. While

---

[22]   For an account of Harlow's work, see Deborah Blum's *Love at Goon Park: Harry Harlow and the Science of Affection*. Interesting in the context of the Bildungsroman and this chapter in particular, Haraway writes that Harlow's articles were "narrated as travel stories" (233). She argues, "To practice science offers the ideal travel experience. Reinforcing the heroic masculinist narrative of self-birthing is the forceps of sadism. It is important to stress that the sadism does not lie, at least originally, in the fact of causing repeated pain to animals in the course of experiments. Rather, the sadism is the organizer of the narrative plot and part of the material apparatus for the cultural production of meanings; sadism is about meanings produced by particular structures of vision, not about pain" (233). In *Primate Visions*, Haraway suggests that Western primatology as a whole has the quality of "travel and quest literature" (251).

[23]   For example, his work on mirror neurons (as they form part of Baron-Cohen's neurological "empathy circuit"): "When a monkey (with a deep electrode in its brain) sees another monkey reaching for an object, cells in the FO increase their electrical activity, and the same cells fire when the monkey reaches for an object itself. ... The mirror system in humans is hard to measure, obviously because it is *unethical to place electrodes into the awake human healthy brain*" (*Zero* 22, 26; italics mine). Interestingly, Baron-Cohen's book includes as appendixes two versions of an empathy test, one for adults and one for children; in the adult test only two out of forty questions pertain to nonhuman animals whereas in the children's test three out of twenty-seven questions ask about other animals.

history and biology refute these stories "which depict our forebears as ferocious, fearless, and free" they continue as potent myths, "as the belief that we can treat the planet any way we want, that humanity will be waging war forever, and that individual freedom takes precedence over community" (25).

These origin-stories, these myths of human superiority, are the negative of our fear of nonhuman nature. They contain the discourse of animality that, I argue, is a two-way fantasy of detachment: we were "beasts" by nature (individualistic, aggressive, and free from restraint) and became humans by culture (social and free to restrain). Our imagined prehistory as one long conquest is the fantasy that we too once had *bestial* access to pleasure ("superior" access—we were superior beasts just as we are now superior as non-beasts), that we could eat and procreate, gorge and fuck, without consideration for others. We imagined we were beasts just as we now imagine we are human. Ethnoracism and dehumanization rely on this story of detachment. It is not simply that we fear that the other thinks (we know "it" thinks), we imagine that the other is "a mind without a heart," a machine without a soul, without *character* (as British anti-Semites used to say), without *proper feeling*, without empathy and so undeserving of sympathy (outside of moral consideration). We imagine this because "we" imagine we want to be minds without hearts. *Cogito ergo sum. We equate freedom with an ability to act without sympathy.*

This way of reading the discourse of animality as a story about empathy, when it has so often been read as a story about reason,[24] parallels Slavoj Žižek's theory of racism as a fantasy about the other's access to pleasure: "the 'other' wants to steal our enjoyment (by ruining our 'way of life') and/or it has access to some secret, perverse enjoyment. In short, what really gets on our nerves, what really bothers us about the 'other', is the peculiar way he organizes his enjoyment" (*Looking Awry* 165). As Tony Myers has it, for Žižek, the "subject of racism, be it the Jew, the Turk, the Algerian, or whoever, is a fantasy figure, someone who embodies the void of the Other" (108); again, in this framework, there are two basic fantasies: "the ethnic 'other' has a strange or privileged access to *jouissance*" and "the ethnic 'other' is trying to steal our *jouissance*" (109).

As agents without empathy (with unrestrained access to pleasure), nonhuman animals and other animalized subjects are constructed not only as Other but, in a sense, as monstrous, as we see in *Candide* (the monkeys), *Frankenstein* (the produced Monster), *Orlando* (the materially fluid, fluidly gendered Orlando), and *A Small Place* (the native who 'inexplicably' blows things up, who mutilates children). On monsters, Žižek writes in *Enjoy Your Symptom!*, "analysis focused on the 'ideological meaning' of monsters overlooks the fact that, previous to signifying something, previous even to serving as an empty vessel of meaning, monsters embody enjoyment *qua* the limit of interpretation, that is to say, *nonmeaning as*

---

[24] The racist discourse of brain size in nineteenth-century anthropology and aspects of the current biological discourse about brain size among species could be read as the negative of this story about empathy.

such" (134). For Žižek, "*you cannot have both meaning and enjoyment*" (134), and so the monstrous is the limit (and subject) of Enlightenment reason:

> In this sense, monsters can be defined precisely as the fantasmatic appearance of the "missing link" between nature and culture: as a kind of "answer of the real" to the Enlightenment's endeavor to find the bridge that links culture to nature, to produce a "man/woman of culture" who would simultaneously conserve his/her unspoiled nature. Therein consists the ambiguity of the Enlightenment: the question of "origins" (origins of language, of culture, of society) which emerged in all its stringency with it, is nothing but the reverse of a fundamental prohibition, the prohibition to probe too deeply into the obscure origins, which betray a fear that by doing so, one might uncover something monstrous ... (136)

This brings us around, again, to the Bildungsroman as humanism's origin story of culture. The Western discourse of animality is a fantasy about others as bestial, as monstrous—as appetite without empathy, pleasure without meaning—and so it is also a fantasy about our "freedom," about the rightness, the "good faith" of not taking their feelings into account. It is a fantasy in bad faith.

The Violence of Freedom

The violence of origins, of constituting them through the other, is the violence of "freedom." Sabine Broeck reads the racism of Enlightenment philosophy[25] as whiteness "refracted" in Kincaid's work: "What Young terms 'desire' (more or less directly linked to sexuality),[26] and what shines in Kincaid's texts as 'mourning and melancholia' of white people, then becomes the signifier for that which is seen as lost in the process of enlightenment: innocence, impulse, spontaneity, eroticism, a connection to nature ... a narcissistic myth that Kincaid's texts mercilessly refract" (825). The narcissistic myth that Broeck describes and Kincaid depicts as white melancholy, the power and privilege of "the Western subject position of whiteness," is part of the fantasy of detachment discussed in the last few pages (and, as I have it earlier in this book, solipsism). Broeck recognizes this myth, this subject-position, as desire: the "*construction* of white rationality ... reveals itself, from Kincaid's perspective, as willfulness, as the *realization of a desire* ..." (838). The power-reifying lament about the Other's proximity to nature (as I have it, our fantasy of the Other's access to bestial pleasure) is a desire for pleasure, as we saw articulated by the tourist of *A Small Place*, but it is more than this too. It is a construction of pleasure as freedom.

Broeck analyzes whiteness as an Enlightenment construction of freedom, a "megasignifier for various European cultures and/or national entities" (827) that proclaims "a human right (over and against feudal restrictions) to knowledge,

---

[25]   Voltaire, notably, is absent here; Broeck discusses Diderot, Rousseau, Kant, Hume, and Hegel.

[26]   See Robert Young's *Colonial Desire: Hybridity in Theory, Culture, and Race.*

self-possession and mastery, in blatant circularity of logic given as an exclusive birthright to white Western Europeans by the Enlightenment scholars because of the self-declared 'beauty' of the Caucasian race" (832–3). She also examines the following passage from *Lucy*, quoted in the last chapter:

> Paul had wanted to show me an old mansion in ruins, formerly the home of a man who had made a great deal of money in the part of the world that I was from, in the sugar industry. ... As we drove along, Paul spoke of the great explorers who had crossed great the seas, not only to find riches, he said, but to feel free, and this search for freedom was part of the whole human situation. Until that moment I had no idea that he had such a hobby – freedom. (129)

While Broeck's reading of this passage as "Kincaid's miniature allegory on European 'great liberals' of the likes of John Locke" (834) is historicized and insightful,[27] it elides the crucial comparison with the other *Other* quite literally caught in the path of Western (and human) history. Broeck does not quote or consider the rest of this scene and paragraph,[28] the two sentences which follow (also quoted in the previous chapter):

> Along the side of the road were dead animals – deer, raccoons, badgers, squirrels – that had been trying to get from one side to the other when fast-moving cars put a stop to them. I tried to put a light note in my voice as I said, "On their way to freedom, some people find riches, some people find death," but I did not succeed. (129)

As I argued in Chapter 5, these dead animals are important. It is through the nonhuman animals in this scene that we see what Lucy later articulates as the "foul deed" of slavery as another cost of people travelling to feel "free" (135).

In *Animal Rites*, Wolfe characterizes this discourse of freedom as the epitome of speciesism. His analysis of philosopher Luc Ferry's *The New Ecological Order*,

---

[27] She argues that Kincaid condenses "white liberalism in Locke's tradition into the image of a 'hobby' of 'freedom' ... Locke's decisive 'progress'—making way for bourgeois self confidence culminating in the enlightened white man of the 19th century— had two components: perfection and possession" (834). As Broeck has it, Locke's argument for freedom is "*freedom as self-possession*," an argument for the white man's self as the foundation of his property (self-ownership) in contradistinction with slavery (834). Broeck then cites James Farr on "Locke's own massive participation in the slave trade" (834).

[28] However, Broeck's analysis of whiteness and women in Kincaid's novels works well: "In her characters of the schoolteacher in *Annie John*, of Mariah in *Lucy*, and Moira in *Autobiography* Kincaid draws particular attention to a peculiar position of white women vis-à-vis white privilege: a vexing ambivalence of mastery and victimization, of denial and want, of loss and parasitism prevents those women—symbolic representations of different, even conflicting subject positions white women have been able to occupy—from addressing their white subjectivity" (830). To me, the passage Broeck chooses to exemplify Moira seems, in particular, to consider whiteness ("her kind") as a kind of biological perfection, in terms reminiscent of the Great Chain of Being (*Autobiography* 208).

an attack on animal rights philosophy and deep ecology, is a case in point. Wolfe notes that Ferry posits freedom as the basis of inclusion in the ethical community, as "the ethical wedge between the human and nonhuman animal," which is untenable because it means "we would be forced to say that the hydrocephalic infant had no interests and rights and could therefore be exploited as pure means (just as laboratory animals are) because it neither embodies nor has the capacity for the liberal 'freedom' that ensures ethical consideration" (36). He exposes this idea as a "question begging concept" (37), revealing what I would call the bad faith inherent in liberal humanism:[29] as with "the term 'democracy,' 'freedom' in his humanist lexicon turns out to be a good deal less free—and a good deal more historically and socially specific—than he would have us believe" (38). It should come as no surprise that Ferry makes a point of arguing (incorrectly) that nonhuman animals have no culture, no "separation from the commandments of nature ... transmitted from *one generation to the next* as a history" (qtd. in Wolfe 41). Aside from the doublethink inherent in positioning culture as freedom, Wolfe notes that "taking account of the ethical relevance of the work of ethologists like Jane Goodall ... *does not mean committing ourselves to naturalism in ethics*" (42).

As a project, *Animal Rites* makes the irrefutable argument that cultural studies relies on a "fundamental repression" of the question of nonhuman subjectivity: "debates in the humanities and social sciences between well-intentioned critics of racism, (hetero)sexism, classism, and all other –isms that are the stock and trade of cultural studies almost always remain locked within an unexamined framework of *speciesism*" (1). And yet, this discourse serves to "mark *any* social other" (7); Wolfe reminds us that,

> as long as it is institutionally taken for granted that it is all right to systematically exploit and kill nonhuman animals simply because of their species, then the humanist discourse of species will always be available for use by some humans against other humans as well, to countenance violence against the social other of *whatever* species—or gender, or race, or class, or sexual difference. (8)

But this is not why we must consider what are called "the animal"—real animal-others; they matter in and to themselves and should matter to us. Wolfe asks, "what does it mean when the aspiration of *human* freedom, extended to all, regardless of race or class or gender, has as its material condition of possibility absolute control over the lives of *nonhuman* others?" (7). For this reason, Wolfe's project is to "disengage the question of a properly postmodern pluralism from the concept

---

[29]   "For how can Ferry locate the basis of ethical consideration in freedom, defined by 'perfectibility, by the capacity to break away from natural or historical determinations' (15), and at the same time praise the way Enlightenment culture recognizes (as in Musil's example) the forces of historical determination to wholly shape one's character?" (Wolfe 37).

of the human ... precisely by taking seriously pluralism's call for attention to embodiment, to the specific materiality and multiplicity of the subject" (9).[30]

The "dead animals" of *Lucy* signify not only the "animalizing" violence of slavery (a "hobby" of freedom) but the violence of the humanist idea of freedom itself, a violence made possible by idea of "the animal" (a two-way fantasy of detachment). Aside from its deployment in a discourse of animality, used to condone or countenance violence against a variety of others, the term "animal" is itself an act of violence. As Derrida argues in *The Animal That Therefore I Am*, from Descartes to the present the whole history of Western philosophy enacts a disavowal of nonhuman animals, a "logic [that] traverses the whole history of humanity," a theorizing ("they made the animal a *theorem*") that "institutes what is proper to man, the relation to itself of a humanity that is above all anxious about, and jealous of, what is proper to it" (14). This description of human anxiety, traced in this book through the Bildungsroman as humanist culture's story about itself, does not so much explain as articulate the conditions of the violence of the idea of "the animal."[31] It is also, as Derrida reminds us, an escalating violence perpetrated on real nonhuman animals: "no one can deny this event—that is the *unprecedented* proportions of this subjection of the animal. ... which some would compare to the worst cases of genocide (there are also animal genocides: the number of species endangered because of man takes one's breath away)" (25–6).

The list of what Western culture denies nonhuman animals is, for Derrida, seemingly inexhaustible—including, of course, subjectivity, language, reason, and culture. These denials are in general, as Derrida has it, "a crime. Not a crime against animality, precisely, but a crime of the first order against the animals, against animals" (48). Derrida's project, like Wolfe's after him, is not so much

---

[30]   Neel Ahuja's "Postcolonial Critique in a Multispecies World" addresses the racist conflation of race and species with a theory of "speciated reason." While he praises the critique of animalization as one strategy for combating the influence of speciated reason in the colonial world, citing Frantz Fanon's *The Wretched of the Earth*, Ahuja argues that animal studies, more specifically the work of Cary Wolfe, continues a deeply problematic conflation of race and species by assimilating "racial discourse into species discourse, flattening out the historical contexts that determine the differential use of animal (and other) figures in the process of racialization ... taking animalization as the generic basis of racism" (557–8). This critique, however, misses the point of the analysis of the discourse of animality/species, which is to foreground and contest speciesism, particularly in and through cultural theory and studies.

[31]   In philosophy and in common sense, "animal" is a word used "to corral a large number of living beings within a single concept. 'The Animal,' they say. And they have given it to themselves, this word, at the same time according themselves, reserving for them, for humans, the right to the word ..." (32) Derrida uses, instead, the term *animot* or *l'animot* to emphasize what "animal/the animal" denies or attempts to erase: multiplicity, difference, and "a response that could be precisely and rigorously distinguished from a reaction; of the right and power to 'respond,' and hence of so many other things that would be proper to man" (32).

to "restore" these capacities or qualities to nonhuman animals, but to question if these things are, as culture knows them, proper to the human and to examine the discourse of animality that makes humanism possible, that is, the discourse of human exceptionalism.

The violence inherent in the idea of "the animal," of naming and reserving names (words) for humans, is the equation of "the animal" with *nonresponse* and lack (81, italics mine). This is the violence that Adorno identifies in Kant's work that, as Derrida has it, leads him to go beyond noting the "affinity between the pretension to transcendentalism and this project of human mastery over nature and over animality" to state that "for an idealist system ... animals virtually play the same role as Jews did for a fascist system. Animals would be the Jews of idealists, who would thus be nothing but virtual fascists. And such a fascism begins whenever one insults an animal, even the animal in man" (103). As I see it, this denial of nonhuman animal response is, fundamentally, a denial of their capacity to respond not primarily to language or logic, but to feeling (*to feel with*, as the root of the word empathy suggests) with sympathy or violence. Again, as Gay Bradshaw notes in her study of elephant PTSD, "*Violence* is a powerful word, and it is not usually employed in the case of animals ... In contrast to the more frequent animal descriptor of aggression, violence includes intent and implies moral violation, attributes typically reserved for the human species" (38, italics in original).

The contradictory ways in which "the animal" is figured (as Derrida has it, everything between absolute goodness and absolute evil, as well as prior to both[32]), the discourse in "all domains that treat the question of the animal," is "the discourse of domination itself. And this domination is exercised as much through an infinite violence, indeed, as through the boundless wrong that we inflict on animals, as through the forms of protest that at bottom share the axioms and founding concepts in whose name the violence is exercised ..." (89). This "criminalization of practical reason ... confirms the waging of a kind of species war and confirms that the man of practical reason remains bestial in his defensive and repressive aggressivity"; it reveals that "bad will, even a perverse malice, inhabits and animates so-called good moral will ..." (101). This is the bad faith of placing nonhuman animals outside moral consideration (Hume) or outside the ethical circuit (Levinas, as Derrida has it). This bring us, through Derrida's analysis of Levinas on dehumanization, back to "monkeys": "The animal remains for Levinas what it will have been for the whole Cartesian-type tradition: a machine that doesn't speak, that doesn't have access to sense, that can at best imitate ... a sort of monkey with 'monkey talk,' precisely what the Nazis sought to reduce their Jewish prisoners to" (117). For Levinas, and others in the Cartesian tradition, "the animal" is not an Other; it is

---

[32] "[T]he dominant discourse of man on the path toward hominization imagines the animal in the most contradictory and generic terms [espèces]: absolute (because natural) goodness, absolute innocence, prior to good and evil, the animal without fault or defect (that would be its superiority as inferiority), but also the animal as absolute evil, cruelty, murderous savagery" (64).

only "a deprivation of humanity" (117). "It" remains the "nonsubject subjected to the human subject" (126).[33] The denial of nonhuman animal empathy, the denial of their capacity to respond to feeling with sympathy or violence, is a denial of their existence in the ethical community, of the politics of our treatment of them and, fundamentally, of the politics of their lives. Again, as Jacques Rancière suggests, what is at stake in the definition of politics is nothing less than the answer to the question: "who is qualified for thinking at all?" (116). To ask who is qualified for politics, what counts as political, is to ask who *counts* full stop. For humanism, the question of who counts is intimately bound up with the question of what counts as culture—a question key to the process of colonization.

In his foreword to the 2004 edition of Fanon's *Wretched of the Earth*, Homi K. Bhabha writes that the "time is right to reread Fanon, according to David Macey, his most brilliant biographer, because 'Fanon was angry,' and without *the basic political instinct of anger* there can be no hope for 'the wretched of the earth [who] are still with us'" (x, italics mine). To put it another way, Fanon affirms anger, affirms certain forms of violence itself, as political. This anger "provides a genealogy of globalization that reaches back to the complex problems of decolonization" (xv). Bhabha sums up the controversy over the role of violence in Fanon's work, most famously articulated by the divergent views expressed in Jean-Paul Sartre's 1961 preface to the text and Hannah Arendt's response to it in *On Violence*: "For Arendt, Fanon's violence leads to the death of politics; for Sartre, it draws the fiery, first breath of human freedom" (xxxvi). For Bhabha, Fanonian violence "confronts the colonial condition of life-in-death" (xxxvi).

The relations between human and nonhuman animals are also life and death relations, and constitute the condition of life-in-death for millions of nonhuman animals; the meat industry is one obvious example. As Achille Mbembe asks in "Necropolitics": "Is the notion of biopower sufficient to account for the contemporary ways in which the political, under the guise of war, of resistance, or of the fight against terror, makes the murder of the enemy its primary and absolute objective?" (12).[34] Though Mbembe here writes about human bodies, he might

---

[33] There are, however, moments in which Derrida's profound analysis of the crime of the idea of "the animal" fails to escape the limitations of humanism. As Haraway suggests in *When Species Meet*, Derrida "does not fall into the trap of making the subaltern speak ... Yet he did not seriously consider an alternative form of engagement either, one that risked knowing something more about cats and how to look back ... He came right to the edge of respect, of the move to *respecere*, but he was sidetracked by his textual canon of Western philosophy and literature and by his own linked worries about being naked in front of his cat" (20).

[34] For all the language of strategy, Ahuja's "Postcolonial Critique in a Multispecies World" limits politics. At the end of his article monkeys, former National Institute of Health laboratory research subjects, are positioned as our "companion travelers" in biopower. "By unwinding the tangled webs of monkeys, human beings, and scientific institutions in visual culture, spatial organization, and popular legends, we can recognize monkeys as companion travelers under imperial biopower" (516). In this formulation lies the problem with Ahuja's

just as well be writing about other animal bodies too—again, as Derrida argues, we have been waging war on "animot" for generations, a "violence, which some would compare to the worst cases of [human] genocide" (26). In fact, Mbembe's rationale for the idea of necropolitics seems a perfect description of the meat industry: because "biopower" does not adequately explain "contemporary forms of subjugation of life to the power of death ... I have put forward the notion of necropolitics and necropower to account for ... the creation of *death-worlds*, new and unique forms of social existence in which vast populations are subjected to conditions of life conferring upon them the status of *living dead*" (39–40, italics in original). They are indeed the wretched of the earth.

   The condition of life-in-death Fanon describes is a process of animalization in a physically and ideologically compartmentalized world, a process rooted in the colonizer's view of other's "bestial" access to pleasure (his "indigence" and "depravity"):

> Values are, in fact, irreversibly poisoned and infected as soon as they come into contact with the colonized. The customs of the colonized, their traditions, their myths, especially their myths, are the very mark of this indigence and innate depravity. This is why we [the colonizers, the West] should place DDT, which destroys parasites, carriers of disease, on the same level as Christianity, which roots out heresy, natural impulses, and evil. ... In plain talk, he [the colonial subject] is reduced to the state of the animal. And consequently, when the colonist speaks of the colonized he uses zoological terms. ... the colonist refers constantly to the bestiary. (7)

Fanon addresses the discourse of animality so early in *Wretched* because it is the rhetorical (and philosophical) foundation of colonial justification. It is, again, a two-way fantasy of detachment: a story about "their" lack of empathy, their unrestrained access to pleasure, and so about the ethical, political correctness of our lack of empathy for them.

---

take on animal studies and, in fact, on nonhuman animals—while it acknowledges that these "webs" constitute or are caught by imperial power, it fails to acknowledge the deeply incommensurate political experiences of monkeys and human animals in this power. Not unlike the canine agility training featured in Haraway's *When Species Meet*, the "rules" are always made by humans and almost always for (some) humans, however modified the game might be by the subaltern, human or nonhuman, in ways we can and cannot see. Ahuja does, however, recognize the latter—the often invisible (and, I would add, invisible-making), organic reach of biopower and resistance to power: "Power in a multispecies world is redirected on scales and in places that often elude perception. The evidence is not always archived on paper or in silicon; it is more often written in the dirt, where our shit combines with bacteria, weeds, and worms to make the soil out of which empires and their discontents grow" (561). It is here in the dirty mix of monkeys and human beings, worms and human shit—reminiscent of the waters surrounding Kincaid's small place, replete with the memory of drowned slaves and tourists' feces—that we may be reminded that the world is (and has always been) multispecies. The radical potential of an ecocultural approach lies in the acknowledgment that it is also *multi*-multicultural.

Interestingly, Arendt's *On Violence* goes to a great deal of trouble to distinguish violence from aggression, to characterize violence as anthropogenic; she stresses the role of tools in human violence ("violence—as distinct from power, force, or strength—always needs implements" [4]) to assert that violence itself is always instrumental (46, 51, 79). Her critique of Fanon (and, to a larger extent, Sartre) is at least in part a critique of biologism, of biological determinism (though she does not use these terms); she refers to the naturalization of violence, the "anthropomorphisms" of the behavioral sciences: "the research results of both the social and natural sciences tend to make violent behavior even more of a 'natural' reaction than we would have been prepared to grant without them" (60). While the "biological justification of violence" is indeed part of "the most pernicious elements in our oldest traditions of political thought" (74), what Arendt lumps together as the biological justification for violence, seemingly all of the behavioral sciences, is (I hope rather obviously by now) *not all* biologism or justification.[35]

Arendt argues that "man" alone is a political being, and what makes him political is "his faculty of action; it enables him to get together with his peers, to act in concert, and to reach out for goals and enterprises ... No other faculty except language, neither reason nor consciousness, distinguishes us so radically from all animal species" (82). "Neither violence nor power is a natural phenomenon," writes Arendt, "they belong to the political realm of human affairs whose essential quality is guaranteed by man's faculty of action, the ability to begin something new" (82). She fears this faculty of action is, for many, frustrated in modernity, leading Pavel Kohout to write that the world needs "'a new example' if 'the next thousand years are not to become an era of *supercivilized monkeys*' or, even worse, of 'man turned into a chicken or rat,'" perhaps ruled by an elite counseled by an "artificial brain" (qtd. in Arendt 83, italics mine). The progress of modernity threatens "devolution"; quite literally, it seems we must avoid acknowledging our animality—as Arendt implies through Kohout, our resemblance to monkeys—to avoid becoming "monkeys."

As I mentioned in the Introduction, Gay Bradshaw's research suggests, quite convincingly, that "man" is *not* the only political animal. Human violence (poaching, "culling," and habitat loss) has led to a collapse of "elephant culture" (*Nature* 807). Bradshaw asks, "How do we respond to the fact that we are causing other species like elephants to ... breakdown? In a way, it's not so much a cognitive or imaginative leap anymore as it is a political one" (Siebert, "An Elephant"). Bradshaw does not simply mean that the fate of nonhuman animals is a political problem for and among humans. Her work implies that human-nonhuman relations (here, human-elephant relations) are themselves political. This is the "leap" she asks us to make—the leap which, as I see it, is a step toward the re*cognition* of the

---

[35] Arendt quotes ethologist Nikolas Tinbergen, behavioral psychologist Erich von Holst, and ethologist Konrad Lorenz, but it seems the last two that she finds objectionable. She does not, however, discuss Kinji Imanishi, the founder of cultural biology (who was also working at this time).

world. As she has it in *Elephants on the Edge*, quoted in the Introduction, elephant violence may be seen as another form of resistance to colonial oppression and global power:

> Much like other cultures that have refused to be absorbed by colonialism, elephants are struggling to survive as an intact society, to retain their elephant-ness, and to resist becoming what modern humanity has tried to make them— passive objects in zoos, circuses, and safari rides, romantic decorations doting the landscape for eager eyes peering from Land Rovers, or data to tantalize our minds and stock in the bank of knowledge. Elephants are, as Archbishop Desmond Tutu wrote about black South Africans living under apartheid, simply asking to live in the land of their birth, where their dignity is acknowledged and respected. (71–2)

Bradshaw's work not only requires the recognition of our relations with elephants (and many other lifeforms) as political, it also suggests that the resistance to the idea of nonhuman animal cultures is not, or not only, intellectual but political.[36]

Again, as Darwin, Freud, and others suggest, empathy, *fellow feeling*, is the glue—the substance—of society. Hume's argument that nonhuman animals cannot participate in human society, and so justice (the treatment due to those with empathy) does not apply to them is, from this perspective, a circular justification. Positioning nonhuman animals as agents without empathy allows us to imagine them outside the ethical community, depriving them of our sympathy, keeping them outside the polis, denying them politics. The humanist discourse of culture, then, is not simply, or only, about the uniqueness of human "reason" (knowledge and ability), but also about the uniqueness of human social feeling.

Garden Politics

Gardening is inextricably interconnected with our perceptions of "nature" and "culture," with ethics and politics—with who counts in our dealings with the world. In a sense, human ethics and politics are equally bound up with gardening, with boundaries real and artificial, with what (who) stays and what (who) goes, with whose needs, fears, and desires count. Early in *My Garden,* Kincaid describes ethics as good domesticity, wishing for

---

[36]   Bradshaw's work recognizes certain acts of elephant violence as collective action. Indeed, it recognizes the existence of not only elephant sympathy but also solidarity. While it may well be true that violence "changes the world, but the most probable change is to a more violent world" (Arendt 80), not all violence is, as Arendt must have known all too well, the same. Elephant violence, and other forms of resistance to oppression, must be considered in cultural and historical context, as a response to the violence of capitalism and colonization.

a recipe for how to make a house a home, a home being a place in which the mystical way of maneuvering through the world in an ethical way, a way universally understood to be ecstatic and universally understood to be the way we would all want it to be, carefully balanced between our own needs and the needs of other people, people we do not know and may never like and can never like, but people all the same who must be considered with the utmost seriousness, the same seriousness with which we consider our own lives. (48)

This construction of ethics as *oikos*, as home, is oddly unrelated to the larger sense of home, the *oikos* of ecology. Instead, Kincaid articulates the ethical connection between the house and the world, the local and global, as a dream of a universal passage through or around difference. Here, difference (human differences, people "we may never like") is a burden on the home, a problem to be considered rather than a quality of the home itself—the key to life. Kincaid's wish for universality instead of the messiness of an ecology, for a "recipe" for right living, a formula for staying clean, makes ethics an impossibility. Here, Kincaid's "tarrying" with the universal seems the wish for a release from the specific, daily gardening of ethical relations. The nesting of self, home, community, and world only makes sense if this world is larger than the human world, if "man" is not the measure of all things, if difference itself (in its *many* different forms) is valued. If nesting itself is a series of interrelations extending to all, in all directions.

Just as *Among Flowers* traffics in dehumanization, *My Garden* presents a self-conscious speciesism. In the following passage, Kincaid encounters a fox in her Vermont garden (almost, but not quite, as Derrida is encountered by his cat):

What to do when the fox looked at me *as if* he was interested in me in *just the way I was interested in him* (who is he, what is he doing standing there just a few steps from my front door, my front door being just a stone's throw from where he/she might be expected to make a den). The fox, after looking at me (for a while, I suppose, though what is a while really?), walked off in that stylish way of all beings *who are confident that the ground on which they put their feet will remain in place*, will remain just where they expect the ground to be. ... And then he disappeared into another part of the wild and I could not follow. (18–19, italics mine)

Though Kincaid announces her ambivalence about her humanity ("I believe I am in the human species, I am mostly ambivalent about this" [18]), she later declares that this same fox, with his prey, is on the opposite side of an insurmountable divide: "the fox is in nature, and in nature things work that way. I am not in nature. I do not find the world furnished like a room, with cushioned seats and rich-colored rugs.[37] To me, the world is cracked, unwhole, not pure, accidental; and the idea of monuments of joy for no reason is very strange" (124).

Here Kincaid characterizes her assertion of human uniqueness as a clear-sighted (one might say "grown-up") recognition of the world's "cracked" nature

---

[37] Alluding to Henry James's *Portrait of a Lady*.

(as opposed to Henry James's painterly natural idealism, his imperialistic vision of nature as a mirror of self, culture, etc.). In this context, wholeness and purity convey a frightening essentialism, the logic of racism, fascism, and dehumanization. The reader has but one choice here, between a foolish, dangerous view of the world and Kincaid's view of human beings as outside nature. The fox in Kincaid's garden is not part of her ethical community, despite being part of her home.

Interestingly, *My Garden* not only discusses the imperial aspects of gardening (from colonial botanical gardens to the "native's" lack of familiarity with indigenous flora and fauna), it considers Empire itself as a form of gardening. Here, Kincaid quotes and comments on the caption of a photo from a book called *The Tropical Garden*:

> "Shortage of labor was never a problem in the maintenance of European features in large colonial gardens; here a team of workers is shown rolling a lawn at the Gymkhana Club in Bombay." ... There must be many ways to have someone be the way you would like them to be; I only know of two with any certainty: You can hold a gun to their head or you can clearly set out before them the thing you would like them to be, and eventually they admire it so much, without even knowing they do so, that they adopt your ways, almost to the point of sickness; they come to believe that your way is their way and would die before they give it up. (141)

Just as gardening is "such an act of will" (111), Empire is an act of bending wills, forming desire, changing (sometimes killing) cultures. In *My Garden*, we see that it has formed Kincaid's desire to create her garden map of the Caribbean, that this desire is *manufactured* (as the title of the book, and all the references to English gardening and garden writers, suggests). It is, in other words, deeply cultural; as Kincaid puts it: "The way you think and feel about gardens and the things growing in them—flowers, vegetables—I can see must depend on where you come from, and I don't mean the difference in opinion and feeling between a person from Spain and a person from England ..." (114).

Returning for a moment to *Among Flowers*, Kincaid writes that the garden is also something "holy," something profound, whose "inevitable boundaries ... must be violated," making the gardener's explanation of her garden the story of a "transgression by a transgressor" (115–16). It is the idea of "an ideal idyll ... between life and death" (169); it is "comfort and beauty" (181). She concludes *Among Flowers* with this last thought about the garden:

> As I walked and observed, each plant, be it tree, shrub, or herbaceous perennial, seemed perfect in its setting or in its sighting. I was in fact looking at Nature, or the thing called so, and I was also looking at a garden. The garden is an invention, the garden is an awareness, a self-consciousness, an artifice. We think and feel that we are making something natural when we make a garden, something that, if come upon unexpectedly, is a pleasure to behold; something that banishes the idea of order and hard work and disappointments and sadness, even as the garden is sometimes made up of nothing but all that. Eden is never

far from the gardener's mind. It is The Garden to which we all refer, whether we know it or not. And it is forever out of reach. ...Vermont, all by itself, should be Eden and gardenworthy enough. But apparently, I do not find it so. I seem to believe that I will find my idyll more a true ideal, only if I can populate it with plants from another side of the world. (188–9)

My Garden makes many of these same points—that all gardens refer back to Eden like it or not (223), that gardening as a form of management is (to some extent) imperialistic, a transgression. Here she describes Eden itself as a garden gone wrong, the garden of a Gardener tired of its upkeep, a garden tired of the Gardener's demands (223). This is the garden of original comfort that Kincaid rejects: "Eden is like that, so rich in comfort, it tempts me to cause discomfort; I am in a state of constant discomfort and I like this state so much I would like to share it" (229). Her hybrid garden in Vermont (a space in America shaped like the Caribbean, grown from England, China, and the other parts of the world her seeds come from) suggests this very state of discomfort: "I mostly worry in the garden, I am mostly vexed in the garden" (19). This postlapsarian garden is the world.

Eden is, for the West, the original origin story. For the Bildungsroman, it is not only the garden to which all others must refer, the origin story of human supremacy over the rest of nature (as Lynn White Jr[38] and so many others have noted), it is also the Ur-plot of human "development," the expulsion from childhood into adulthood, from nature into culture. It is, in this way, also the framework for the humanist origin story of culture—of its creation of itself from the soil of nature.[39] Explicitly, the genre of the Bildungsroman is the story of the individual coming into culture ("coming of age") but, as I have argued in this book, it is fundamentally culture's story of coming out of and apart from nature, just as gardening is a cultivation of nature, a narrative of bringing nature into culture. As Kincaid writes, "An integral part of a gardener's personality—indeed, a substantial amount of a gardener's world—is made up of the sentiment expressed by the two words 'To Come'" (*My Garden* 85). The garden is a Bildungsroman, and the Bildungsroman is a garden.

The idea of Eden is, one might argue, at the heart of the genre's revival by postcolonial novelists. While Kincaid suggests that it may be a coincidence that so many postcolonial writers have produced coming–of–age novels,[40] this seems far from the case. Again, as MacDonald-Smythe argues, these authors take on the genre to challenge and recast the Western (white male) model of development with which it is identified, to render it "more applicable to the discursive formulation of the marginal subject" (29). The postcolonial interest in the Bildungsroman seems even less like a coincidence given the genre's tendency toward an Orientalizing

---

[38]   See "The Historical Roots of Our Ecological Crisis."

[39]   As Haraway writes of mainstream science, quoted in Chapter 5: "the ripening of the human from the soil of the animal" (*Primate* 11).

[40]   "*Nervous Conditions* ... is a coming-of-age novel (and really, most people who come from the far parts of the world who write books at some point write about their childhood—I believe it is a coincidence)" (*My Garden* 116).

longing or search for a utopia, as paradise or the power to make one (of course, some of these authors come from those very "far parts of the world"[41] figured in or imagined by traditional examples of the genre as undifferentiated or specific Others[42]). For example, we see the discovery of El Dorado in *Candide*, the Edenic power of creation (possibility of eternal life) in *Frankenstein*, and the escape to Constantinople in *Orlando*. In all three cases, of course, this Orientalizing impulse (if not the narrative of origins itself) is not only critiqued but rejected, a movement central to each work. In *Frankenstein*, the idea of the "Orient" itself takes such a turn; Safie, the "Arabian" from Constantinople, is rescued from the fate of "the female followers of Mahomet" by her union with Felix De Lacey, which allows her to remain "in a country where women were allowed to take a rank in society" (99). Here, the West seems *almost* utopian—as Safie has it, "enchanting"—compared to the East (the very same East, Constantinople, where Orlando becomes a woman), but Shelley and the readers of the novel know there is no such paradise.[43] As we have seen, the garden itself figures as the place of origin in these Bildungsromane: Candide is cast out of his childhood castle-garden and ends up in a garden community of his making; Victor Frankenstein leaves the pastoral domesticity of his childhood and finds himself, in the end, in the wilderness of oceanic exploration; and Orlando quits his grand manor-house garden, with its emblematic Oak Tree, only to return to it, transformed through *her* travels. The journey out of the originary garden is, explicitly, the journey out of "nature" and into culture, but it is also (negatively) the journey into Western culture's fantasy about nature. It is, I argue, a two-way fantasy of detachment, a fantasy (and anxiety) about desire and empathy figured (and elicited) by the monkey-lovers in *Candide*, the Monster in *Frankenstein*, the transmaterial (and transgendered) body in *Orlando*, and the tourist's paradise in *A Small Place*.

In the humanist origin story of culture, culture may seem like a garden, a cultivation of nature, yet it knows (though it says, sometimes insists, the reverse) that culture is wild—the garden itself, as a cultural work, is wild! Even human language—the ability which, above of all others, has been (and still is) used most often to separate human from other animals—is, as the poet Gary Snyder has it, "wild."[44] But what, one might ask, does it mean to be wild? What does it mean that our very practice of telling stories is wild (we are *homo narrator* perhaps more than *homo sapiens*)? It means that culture, which some are fond of calling "second nature," is no *second* nature. It *is* nature (and not only ours). Like Candide's chosen garden, this is no mere metaphor. Human consciousness, socially mediated

---

[41]   See previous footnote.

[42]   Interestingly, the mythical Shangri-La is not named but alluded to by way of imagery in *Among Flowers*, as "the place from which skies were made and then dispatched to other parts of the globe" (164).

[43]   Temma Berg argued this quite convincingly in her talk, "*Frankenstein*: Engendering a Text, Embodying a Text," at East Carolina University on November 2, 2010.

[44]   See *The Practice of the Wild: Essays*. Also, see the emerging field of biosemiotics.

and embedded, is no secondary, meta, or unique quality of the world. It is one form of something that, as it turns out, may be relatively common among social animals (indeed, it may be more common than not). Cultural biology tells us that human cultures are only one of the many kinds of culture in nature; it contends and confirms that which we, like the tourist, already know—what human cultures most likely have always already known.

Culture is ordinary; the world is mediated by multiple cultures in nature. A web of complex mediation, the world is *immediate*. Not in the old humanist sense of the word, but in a way that suggests how deeply and fundamentally we are all materially, politically entangled. In this context, Žižek's clever, fashionable charge against ecological thought as "the crucial field of ideology today"[45] rings false. While Žižek is certainly right that the view of nature as a static harmony is a form of mystification (though one must point out that ecologists have known this for a very long time), ecology itself is not simply another origin story as he claims here:

> It's really the implicit premise of ecology that the existing world is the best possible world, in the sense that it is a balanced world undisturbed by human hubris. So why do I find this problematic? Because I think that this notion of nature, nature as a harmonious, organic, balanced, reproducing almost living organism, which is then disturbed, perturbed, derailed through human hubris, technological exploitation, and so on, is, I think, a secular version of the religious story with a fall. The answer should be not that there is no fall, that we are a part of nature, but on the contrary that *there is no nature*. (*Examined* 159, italics in original)

Ultimately, Žižek's twenty-first century ideological analysis returns to (in fact, embraces) a view of the world as "red in tooth and claw." His claim, that nature is in fact "a series of unimaginable catastrophes" (159), is as ridiculously narrow and partial a view as the one he critiques in the previous passage; both "best possible world" and "series of unimaginable catastrophes" echo the two poles of *Candide*, between (or, more accurately, beyond) which Candide wisely forms his garden. Indeed, contemporary ecological science makes neither claim.

Žižek argues that the ecological narrative is false because it enables the logic of "disavowal" (161), the logic that allows us to continue our consumption of fossil fuels, for example, in the face of climate change. It should be no surprise, then, that Žižek recommends we embrace technology and "cut off even more our roots in nature" (161). His claim seems simple: "ecology is the greatest threat [to us], but to confront it properly we must get rid of this notion of nature ... We should become more artificial" (161–2). For Žižek, human existence itself is, at this point in history, artificial, and we should openly acknowledge and embrace our anthropocentrism and fight for the future we want: "we will still fight pollution, dangers, and so on, but I am tempted to say that we do it as open warfare" (181).

---

[45] *Examined Life* 156.

This argument against nature should not be confused with Derrida's argument against "the animal." In fact, Žižek's vision of an unnatural future sharpens the contradiction, as one might say. While Derrida argues against the humanist narrative of supremacy exemplified by the use of the phrase "the animal," the corralling of all nonhuman animals into one undifferentiated category of inferiority, and the war on real animals of which this ideology is a part, Žižek's unnatural war against ecological threat is a war against threats to humanity, an anthropocentric war, one that exhorts us to embrace the costs of our desires openly. In other words, it is one that purposefully, intentionally does not take other creatures into account (save their benefit to us).

Again, recognizing the existence of other animal cultures—and, in so doing, rejecting various ideologies of nature, including that of human supremacy— benefits many species, challenging structures of power that oppress both human and nonhuman animals. By considering the Bildungsroman as humanism's origin-story of culture, as a claim about our radical uniqueness, our separation from the rest of the living world, we are able to see the ways in which even humanism, as Freud says of religion, "cannot keep its promise" (*Civilization* 36). In the breach of this promise of human superiority lies the hope of humanity and many of the other inhabitants of the earth. A new, multispecies multiculturalism may intervene in forms of oppression that have long functioned by excluding some—human and nonhuman—from the realm of culture. It is an intervention that argues that to think politically, to create a more worldly politics, responsible to similarity as well as difference, we must contest the humanist ideology of culture.

Bibliography

Abel, Elizabeth, Marianne Hirsch, and Elizabeth Langland, eds. Introduction. *The Voyage In, Fictions of Female Development*. Hanover and London: UP of New England, 1983. Print.

Abram, David. *The Spell of the Sensuous*. New York: Vintage, 1996. Print.

Adorno, Theodor. *Aesthetic Theory*. Trans. Robert Hullot-Kentor. Ed. Hullot-Kentor. Minneapolis: U of Minnesota P, 1997. Print.

———. *Critical Models*. Trans. Henry W. Pickford. New York: Columbia UP, 1998. Print.

———. *Metaphysics*. Trans. Rolf Tiedemann. Stanford: Stanford UP, 2000. Print.

———. *Minima Moralia*. Trans. E.F.N. Jephcott. London: Verso, 1974. Print.

———. *Negative Dialectics*. Trans. E.B. Ashton. New York: Continuum, 1973. Print.

———. *Notes to Literature*. Trans. Shierry Weber Nicholsen. 2 vols. New York: Columbia UP, 1991. Print.

———. *Prisms*. Trans. Samuel and Shierry Weber. Cambridge: MIT P, 1967. Print.

Ahuja, Neel. "Postcolonial Critique in a Multispecies World" *Modern Language Association of America* 124.2 (2009): 556–63. Print.

Alaimo, Stacy. *Bodily Natures: Science, Environment, and the Material Self*. Bloomington: Indiana UP, 2010. Print.

———. "Eluding Capture: The Science, Culture, and Pleasure of 'Queer' Animals." *Queer Ecologies: Sex, Nature, Politics, Desire*. Ed. Catriona Mortimer-Sandilands, Bruce Erickson, and Stacy Alaimo. Bloomington: Indiana UP, 2010. 51–72. Print.

Anderson, Roland, and Jennifer Mather. "Personalities in Octopuses." *Journal of Comparative Psychology* (1993): 336–40. Print.

Antigua. Ministry of Tourism and Civil Aviation. *Tourist Arrivals by Purpose of Visit and Month*. St. John's, 2008. Web.

Arendt, Hannah. *On Violence*. 1970. Orlando and New York: Harvest, 1969. Print.

Badmington, Neil. "Theorizing Posthumanism." *Cultural Critique* 53 (2003): 10–27. Print.

Baldanza, Frank. "*Orlando* and the Sackvilles." *Modern Language Association of America* 70.1 (1955): 274–9. Print.

Baldick, Chris. *In Frankenstein's Shadow: Myth, Monstrosity, and Nineteenth-century Writing*. Oxford: Oxford UP, 1987. Print.

———. "The Politics of Monstrosity." *Frankenstein: Mary Shelley*. Ed. Fred Botting. New York: St. Martin's, 1995. Print.

Barash, David P., and Nanelle Barash. *Madam Bovary's Ovaries: A Darwinian Look at Literature*. New York: Delacourt, 2005. Print.

Baron-Cohen, Simon. *Zero Degrees of Empathy: A New Theory of Human Cruelty.* London and New York: Allen Lane, 2011. Print.

Bate, Jonathan. "Living With the Weather." *Studies in Romanticism* 35.3 (1996): 431–47. Print.

———. *Romantic Ecology.* London: Routledge, 1991. Print.

———. *The Song of the Earth.* London: Picador, 2000. Print.

Bekoff, Marc. *The Emotional Lives of Animals.* Novato, CA: New World Library, 2007. Print.

Bell, Quentin. *Virginia Woolf: A Biography.* New York and London: Harcourt Brace, 1972. Print.

Bennett, Tony, Lawrence Grossberg, and Meaghan Morri, eds. *New Keywords: A Revised Vocabulary of Culture and Society.* Oxford: Blackwell, 2005. Print.

Berreby, David. "Deceit of the Raven." *New York Times Magazine* 4 Sept. 2005: 20–22. Print.

Besterman, Theodore. *Voltaire.* Banbury: Cheney and Sons, 1969. Print.

Bhabha, Homi K. *The Location of Culture.* London and New York: Routledge, 1994. Print.

———. Foreword. *Wretched of the Earth.* New York: Grove, 2004. vii–xli. Print.

Black, Stephanie. "About the Film." *Life and Debt.* Tuff Gong Pictures, 2001. 20 Sept. 2009. Web.

Blackburn, Robin. *The Making of New World Slavery: From the Baroque to the Modern, 1492–1800.* New York: Verso, 1997. Print.

Blakeslee, Sandra. "Minds of Their Own: Birds Gain Respect." *New York Times* 1 Feb. 2005: D1. Print.

Bloom, Harold, ed. *Romanticism and Consciousness.* New York: Norton, 1970. Print.

———. *The Visionary Company.* Ithaca: Cornell UP, 1971. Print.

Bonetti, Kay, Greg Michalson, Speer Morgan, Jo Sapp, and Sam Stowers. "Jamaica Kincaid." Conversations with American Novelists: The Best Interviews from The Missouri Review and the American Audio Prose Library. Columbia and London: U of Missouri P, 1997. 26–38. Print.

Botting, Fred. "Reflections of Excess: *Frankenstein*, the French Revolution, and Monstrosity." *Frankenstein.* Ed. Johanna M. Smith. Boston: Bedford. 2000. 435–49. Print.

Bowie, Malcolm. "Freud and the Art of Biography." *Mapping Lives: The Uses of Biography.* Ed. Peter France and William St. Clair. Oxford: Oxford UP, 2002. 177–92. Print.

Bradbury, Malcolm, and James McFarlane, eds. *Modernism: 1890–1930.* New York: Penguin, 1976. Print.

———. *The Social Context of Modern English Literature.* Oxford: Blackwell, 1971. Print.

Bradshaw, Gay. "An Ape Among Many: Animal Co-Authorship and Trans-species Epistemic Authority." *Configurations* 18.1–2 (2010): 15–30. Print.

———. *Elephants on the Edge: What Animals Teach Us about Humanity.* New Haven and London: Yale, 2009. Print.

Bradshaw, Gay, A.N. Schore, J.L. Brown, J.H. Poole, and C.J. Moss. "Elephant Breakdown." *Nature* 433.7028 (2005): 807. Print.

Braziel, Jana Evans. "Daffodils, Rhizomes, Migrations: Narrative Coming of Age in the Diasporic Writings of Edwidge Danticat and Jamaica Kincaid." *Meridians: Feminism, Race, and Transnationalism* 3.2 (2003): 110–31. Print.

Broeck, Sabine. "When Light Becomes White: Reading Enlightenment through Jamaica Kincaid's Writing." *Callaloo* 25.3 (2002): 821–43. Print.

Brooks, Peter. "What is a Monster? (According to *Frankenstein*)." *Frankenstein/ Mary Shelley*. Ed. Fred Botting. New York: St. Martins, 81–106. Print.

Brown, Marshall. "Romanticism and Enlightenment." *The Cambridge Companion to British Romanticism*. Cambridge: Cambridge UP, 1993. 25–47. Print.

Brown, Patricia Lee. "For Penguins, a New Will to Swim." *San Francisco Chronicle* 16 Jan. 2003: A3. Print.

Buell, Lawrence. *The Future of Ecocriticism*. Malden and Oxford: Blackwell, 2005. Print.

———. *Writing for an Endangered World*. Cambridge: Belknap, 2001. Print.

Bugnyar, Thomas. "Leading a Conspecific Away from Food in Ravens (Corvus corax)." *Animal Cognition* (2004): 69–76. Print.

Butler, Marilyn. "*Frankenstein* and Radical Science." *Frankenstein: A Norton Critical Edition*. Ed. J. Paul Hunter. New York: Norton, 1996. 302–13. Print.

———. Introduction. *Frankenstein or, The Modern Prometheus (The 1818 Text)*. Ed. Marilyn Butler. 1994. Oxford and New York: Oxford UP, 1998. ix–li. Print.

Caldwell, Janis McLarren. "Sympathy and Science in *Frankenstein*." *The Ethics in Literature*. Ed. Andrew Hadfield, Dominic Rainsford, and Tim Woods. Houndsmills: Macmillan. 1998. 262–74. Print.

Cantor, Paul. *Creature and Creator: Mythmaking and English Romanticism*. New York: Cambridge UP, 1984. Print.

Cantrell, Carol H. "'The Locus of Compossibility;' Virginia Woolf, Modernism, and Place." *Interdisciplinary Studies in Literature and Environment* 5.2 (1998): 25–40. Print.

Chambers, Douglas. *The Planters of the English Landscape Garden: Botany, Trees and the Georgics*. New Haven, Yale UP, 1993. Print.

"Chimpanzees 'Hunt Using Spears'." *BBC.co.uk*. British Broadcasting Corporation, 22 Feb. 2007. Web.

Clubbe, John. "The Tempest-Toss'd Summer of 1816: Mary Shelley's *Frankenstein*." *Byron Journal* 19 (1991): 26–40. Print.

Collett, Anne. "A Snake in the Garden of the *New Yorker*?: An Analysis of the Disruptive Function of Jamaica Kincaid's Gardening Column." *Missions of Interdependence: A Literary Directory*. Ed. Gerhard Stilz. Amsterdam, Netherlands: Rodopi, 2002. 95–106. Print.

Cudjoe, Selwyn R. "Jamaica Kincaid and the Modernist Project: An Interview." *Caribbean Women Writers: Essays from the First International Conference*. Amherst: U of Massachusetts P, 1990. 215–32. Print.

Curtis, Valerie, and Alison Jolly. "Sick as a Parrot." *London Review of Books* 10 July 2003: 29. Print.

Darwin, Charles. *The Expression of the Emotions in Man and Animals.* 1872. Chicago and London: U of Chicago P, 1965. Print.

Da Silva, N. Takei. *Modernism and Virginia Woolf.* Windsor: Windsor Publications, 1990. Print.

Dawson, P.M.S. "'The Empire of Man': Shelley and Ecology." *Shelley: Poet and Legislator of the World.* Ed. Betty T. Bennett. Baltimore: John Hopkins UP, 1996. Print.

———. *The Unacknowledged Legislator: Shelley and Politics.* Oxford: Clarendon, 1980. Print.

DeLoughrey, Elizabeth M., Renée K. Gosson, and George B. Handley. Introduction. *Caribbean Literature and the Environment: Between Nature and Culture.* Charlottesville and London: Virginia UP, 2005. 1–30. Print.

Derrida, Jacques. *The Animal That Therefore I Am.* Trans. David Wills. Ed. Marie-Louis Mallet. New York: Fordham UP, 2008. Print.

De Waal, Frans. *Age of Empathy: Nature's Lessons for a Kinder Society.* New York: Three Rivers, 2009. Print.

———. *The Ape and the Sushi Master: Cultural Reflections by a Primatologist.* New York: Basic Books, 2001. Print.

De Waal, Frans, and Kristin E. Bonnie. "In Tune with Others: The Social Side of Primate Culture." *The Question of Animal Culture.* Ed. Kevin N. Laland and Bennett G. Galef. Cambridge and London: Harvard UP, 2009. 19–40. Print.

Dixit, Kunda. "Less Food, More Mouths to Feed: New Report Warns of an Impending Food Emergency in Nepal." *The Nepali Times* 7 Aug. 2009. Web. 24 Feb. 2010.

Donnell, Allison. *The Routledge Reader in Caribbean Literature.* Ed. Allison Donnell and Sarah Lawson Welsh. New York and London: Routledge, 1996. Print.

———. "She Ties Her Tongue: The Problems of Cultural Paralysis in Postcolonial Criticism." *Jamaica Kincaid: Modern Critical Views.* Ed Harold Bloom. Philadelphia: Chelsea House, 1998. 37–49.

———. "When Daughters Defy: Jamaica Kincaid's Fiction." *Women: A Cultural Review.* 4.1 (1993): 18–26. Print.

Donovan, Josephine. "Ecofeminist Literary Criticism: Reading the Orange." *Ecofeminist Literary Criticism.* Ed. Greta Gaard and Patrick D. Murphy. Urbana and Chicago: U of Illinois P, 1998. 74–96. Print.

Eagleton, Terry. *The Idea of Culture.* Oxford and Malden: Blackwell, 2000. Print.

———. Introduction. *Ideology.* Ed. Terry Eagleton. London and New York: Longman, 1994. 1–20. Print.

———. *Literary Theory.* Minneapolis: U of Minnesota P, 1983. Print.

Edel, Leon. *Writing Lives: Principia Biographica.* New York: Norton, 1984. Print.

Edlmair, Barbara. "Rewriting History: Alternative Versions of the Caribbean Past in Michelle Cliff, Rosario Ferré, Jamaica Kincaid, and Daniel Maximin." *Austrian Studies in English* 84. Wein: Braumuller, 1999. Print.

Edwards, Justin D. *Understanding Jamaica Kincaid*. Columbia: U of South Carolina P, 2007. Print.

Eisley, Loren. *The Invisible Pyramid*. Lincoln: Bison-U of Nebraska P, 1998. Print.

Endo, Paul. "'Mont Blanc,' Silence, and the Sublime." *English Studies in Canada* 21.3 (1995): 283–300. Print.

Evernden, Neil. "Beyond Ecology: Self, Place, and the Pathetic Fallacy." *The Ecocriticism Reader*. Ed. Cheryll Glotfelty and Harold Fromm. Athens: U of Georgia P, 1996. 92–104. Print.

Ewert, Jeanne C. "'Great Plant Appropriators' and Acquisitive Gardeners: Jamaica Kincaid's Ambivalent Garden (Book)." *Jamaica Kincaid and Caribbean Double Crossings*. Newark: U of Delaware P, 2006. 113–26. Print.

Eysteinsson, Astradur. *The Concept of Modernism*. Ithaca: Cornell UP, 1990. Print.

Eze, Emmanuel Chukwudi. *Race and Enlightenment: A Reader*. Cambridge and Oxford: Blackwell, 1997. Print.

Fanon, Frantz. *The Wretched of the Earth*. New York: Grove, 2004. Print.

Ferguson, Moira. *Colonialism and Gender from Mary Wollstonecraft to Jamaica Kincaid*. New York: Columbia UP, 1993. Print.

———. *Jamaica Kincaid: Where the Land Meets the Body*. Charlottesville and London: U of Virginia P, 1994. Print.

———. "*Lucy* and the Mark of the Colonizer." *Jamaica Kincaid: Modern Critical Views*. Ed. Harold Bloom. Philadelphia: Chelsea House, 1998. 51–69. Print.

Ferres, Kay. "Gender, Biography, and the Public Sphere." *Mapping Lives: The Uses of Biography*. Ed. Peter France and William St. Clair. Oxford: Oxford UP, 2002. 303–19. Print.

Foucault, Michel. *The Order of Things: An Archeology of the Human Sciences*. 1970. New York: Vintage, 1994. Print.

Fragaszy, Dorothy M., and Susan Perry. *The Biology of Traditions: Models and Evidence*. Cambridge: Cambridge UP, 2003. Print.

Fraiman, Susan. *Unbecoming Women: British Women Writers and The Novel of Development*. New York and Oxford: Columbia UP, 1993. Print.

Freeman, Barbara Claire. "*Frankenstein* with Kant: A Theory of Monstrosity or the Monstrosity of Theory." *Frankenstein/Mary Shelley*. Ed. Fred Botting. New York: St. Martin's, 1995. 191–205. Print.

Freud, Sigmund. *Civilization and its Discontents*. Trans. James Strachey. 1930. New York and London: Norton, 1961. Print.

———. *The Joke and its Relation to the Unconscious*. Trans. Joyce Crick. 1905. New York and London: Penguin, 2003. Print.

Fukuyama, Francis. *Our Posthuman Future: Consequences of the Biotechnology Revolution*. New York: Farrar, Straus, Giroux, 2002. Print.

Galef, Bennett. "The Question of Animal Culture." *Human Nature* 3.2 (1992): 157–78. Print.

Gay, Peter. *Voltaire's Politics: The Poet as Realist*. New Haven and London: Yale UP, 1988. Print.

Gilbert, Sandra M. Introduction. *Orlando; A Biography*. London: Penguin, 1993. Print.

Gilbert, Sandra M., and Susan Gubar. *The Madwoman in the Attic*. New Haven and London: Yale, 1979. Print.

Glissant, Édouard. *The Caribbean Discourse: Selected Essays*. Charlottesville and London: U of Virginia P, 1999. Print.

Glotfelty, Cheryll. Introduction. *The Ecocriticism Reader*. Ed. Cheryll Glotfelty and Harold Fromm. Athens: Georgia UP, 1996. Print.

Goldberg, David Theo. *Racist Culture: Philosophy and the Politics of Meaning*. Oxford: Blackwell, 1993. Print.

Goodall, Jane. *The Chimpanzees of Gombe: Patterns and Behavior*. Cambridge: Harvard UP, 1986. Print.

Gordon, Daniel. Introduction: "Postmodernism and the French Enlightenment." Ed. Daniel Gordon. *Postmodernism and the Enlightenment*. New York: Routledge, 2001. 1–6. Print.

———. Introduction. *Candide*. Trans. Daniel Gordon. Boston: Bedford/St. Martin, 1999. 1–30. Print.

———. "On the Supposed Obsolescence of the French Enlightenment." *Postmodernism and the Enlightenment*. Ed. Daniel Gordon. New York: Routledge, 2001. 201–21. Print.

Gosling, Sam, and John P. Oliver. "Personality Dimensions in Nonhuman Animals: A Cross-Species Review." *Current Directions in Psychological Science*. 8.3 (1999): 69–75. Print.

Grosz, Elizabeth. *Time Travels: Feminism, Nature, Power*. Durham and London: Duke UP, 2005. Print.

Haraway, Donna. *Primate Visions: Gender, Race, and Nature in the World of Modern Science*. New York and London: Routledge, 1989. Print.

———. *Simians, Cyborgs, and Women*. Free Association Books: London, 1991. Print.

———. *When Species Meet*. Minneapolis and London: U of Minnesota P, 2008. Print.

Hardin, James, ed. Introduction. *Reflection and Action: Essays on the Bildungsroman*. Columbia: U of South Carolina P, 1991.Print.

Havens, George R. "Background of *Candide*." *Voltaire's* Candide *and the Critics*. Ed. Milton P. Foster. Belmont: Wadsworth, 1962. Print.

Hayles, N. Katherine. *How We Became Posthuman: Virtual Bodies in Cybernetics, Literature, and Informatics*. Chicago and London: U of Chicago P, 1999. Print.

Henry, Patrick. *Voltaire and Camus: The Limits of Reason and the Awareness of Absurdity*. Banbury: Cheney and Sons LTD, 1975. Print.

Hoffmann, Charles G. "Fact and Fantasy in *Orlando*: Virginia Woolf's Manuscript Revisions." *Texas Studies in Literature and Language* 10 (1968): 435–44. Print.

Holmes, Richard. *Footsteps: Adventures of a Romantic Biographer*. London: Hodder and Stoughton, 1985. Print.

Horkheimer, Max, and Theodor Adorno. *Dialectic of Enlightenment.* Trans. Edmund Jephcott. Ed. Gunzelin Schmid Noerr. Stanford: Stanford UP, 2002. Print.

Hoving, Isabel. *In Praise of New Travelers: Reading Caribbean Migrant Women's Writing.* Stanford: Stanford UP, 2001. Print.

———. "Remaining Where You Are: Kincaid and Glissant on Space and Knowledge." *Mobilizing Place, Placing Mobility: The Politics of Representation in a Globalized World.* Ed. Ginette Verstraete and Tim Cresswell. Amsterdam: Rodopi, 2002. 125–40. Print.

Hussey, Mark. *Virginia Woolf A to Z: A Comprehensive Reference for Students, Teachers, and Common Readers to her Life, Work, and Critical Reception.* New York: Facts on File, 1995. Print.

Hussey, Mark and Vara Neverow, eds. *Virginia Woolf: Emerging Perspectives: Selected Papers from the Third Annual Conference on Virginia Woolf.* Lincoln University, Jefferson City, MO. June 10–13, 1993. New York: Pace UP, 1994. Print.

Jagger, Alison M., ed. *Feminist Politics and Human Nature.* Sussex and Totowa: Rowman and Allanheld, 1983. Print.

Jay, Martin. *Marxism and Totality: The Adventures of a Concept from Lukács to Habermas.* Berkeley: U of California P, 1984. Print.

Jordanova, Ludmilla. "Melancholy Reflection: Constructing an Identity for Unveilers of Nature." *Frankenstein, Creation and Monstrosity.* Ed. Stephen Bann. London: Reaktion, 1994: 60–76. Print.

Jouve, Nicole Ward. "Virginia Woolf and Psychoanalysis." *The Cambridge Companion to Virginia Woolf.* Ed. Susan Sellers. Cambridge: Cambridge UP, 2000. 245–72. Print.

Kaplan, Caren. "Reconfigurations of Geography and Historical Narrative: A Review Essay." *Public Culture* 3.1 (1990) 25 32. Print.

Kendrick, Thomas D. *The Lisbon Earthquake.* London: Methuen, 1956. Print.

Khasnabish, Ashmita. "The Theme of Globalization in Kincaid's *Among Flowers.*" *Anthurium* 7.1. 1–2. 2009. Web. 1 Mar. 2010.

Kincaid, Jamaica. *Among Flowers: A Walk in the Himalaya.* Washington D.C.: National Geographic, 2005. Print.

———. *Annie John.* New York: Farrar, Straus, Giroux, 1983. Print.

———. *Autobiography of My Mother.* New York: Plume, 1997. Print.

———. *Lucy.* New York: Farrar, Straus, Giroux, 1990. Print.

———. *My Garden (Book):.* New York: Farrar, Straus, Giroux, 1999. Print.

———. *A Small Place.* New York: Farrar, Straus and Giroux, 1988. Print.

Kolbert, Elizabeth. *Field Notes from a Catastrophe: Man, Nature, and Climate Change.* New York: Bloomsbury, 2006. Print.

Kroeber, Karl. *Ecological Literary Criticism.* New York: Columbia UP, 1994. Print.

Laland, Kevin N., and Gillian R. Brown. *Sense and Nonsense: Evolutionary Perspectives on Human Behavior.* Oxford: Oxford UP, 2002. Print.

Laland, Kevin N., and Bennett G. Galef, eds. *The Question of Animal Culture.* Cambridge and London: Harvard UP, 2009. Print.

Laland, Kevin N., and William Hoppitt. "Do Animals Have Culture?" *Evolutionary Anthropology* 12.3 (2003): 150–59. Print.

Lawrence, D.H. *Phoenix.* Harmondsworth: Penguin, 1985. Print.

Lawrence, Karen R. "Orlando's Voyage Out." *Modern Fiction Studies* 38.1 (1992): 253–77. Print.

Lawrence, William. *An Introduction to Comparative Anatomy and Physiology; Being the Two Introductory Lectures Delivered at the Royal College of Surgeons, on the 21st and 25th of March, 1816.* London: J. Callow, 1816. Print.

Lee, Hermione. *Virginia Woolf.* London: Chatto & Windus, 1996. Print.

Lehman, John. *Virginia Woolf and Her World.* London: Harvest/Harcourt, 1975. Print.

Leopold, Aldo. *A Sand County Almanac.* London, Oxford, New York: Oxford UP, 1949. Print.

Levine, George, U.C. Knoepflmacher, and Peter Dale Scott, eds. *The Endurance of* Frankenstein: *Essays on Mary Shelley's Novel.* Berkeley: U of California P, 1979. Print.

Levins, Richard, and Richard Lewontin. *The Dialectical Biologist.* Cambridge: Harvard UP, 1985. Print.

Lima, Maria Helena. "Decolonizing Genre: Jamaica Kincaid and the Bildungsroman." *Genre* 26 (1993): 431–59. Print.

———. "Imaginary Homelands in Jamaica Kincaid's Narratives of Development." *Callaloo* 25.3 (2002): 857–67. Print.

Lipking, Lawrence. "*Frankenstein,* the True Story: or, Rousseau Judges Jean-Jacques." *Mary Shelley, Frankenstein: The 1818 Text, Contexts, Nineteenth-Century Responses, Criticism.* Ed. J.P. Hunter. New York: W.W. Norton, 1995. Print.

Lokke, Kari Elise. "*Orlando* and Incandescence: Virginia Woolf's Comic Sublime." *Modern Fiction Studies* 38.1 (1992): 235–52. Print.

Love, Glen A. *Practical Ecocriticism: Literature, Biology, and the Environment.* Richmond: U of Virginia P, 2003. Print.

Ludwig, Emil. *Genius and Character.* Trans. Kenneth Burke. London: Harcourt, Brace and Company, 1927. Print.

MacDonald-Smythe, Antonia. *Making Homes in the West/Indies: Constructions of Subjectivity in the Writings of Michelle Cliff and Jamaica Kincaid.* New York and London: Garland, 2001. Print.

Macherey, Pierre. *A Theory of Literary Production.* 1966. London: Routledge, 1986. Print.

Marchi, Dudley M. "Virginia Woolf Crossing the Borders of History, Culture, and Gender: the Case of Montaigne, Pater, and Gournay." *Comparative Literature Studies* 34.1 (1997): 1–30. Print.

Marshall, David. *The Surprising Effects of Sympathy: Marivaux, Diderot, Rousseau, and Mary Shelley.* Chicago: U of Chicago P, 1988. Print.

Marx, Karl. *Capital*. Trans. Ben Fowkes. Vol. 1. 1976. New York and London: Penguin, 1979. Print.

―――. *Marx: Early Political Writings*. Ed. Joseph O'Malley. Cambridge: Cambridge UP, 1994. Print.

Marzluff, John M., and Tony Angell. *In the Company of Crows and Ravens*. New Haven and London: Yale UP, 2005. Print.

Mbembe, Achille. "Necropolitics." *Public Culture* 15.1 (2003): 11–40. Print.

McFarlane, James. "The Mind of Modernism." *Modernism 1890–1930*. Ed. Malcolm Bradbury and James McFarlane. London: Penguin, 1976, 71–93. Print.

McGann, Jerome J. *The Romantic Ideology*. Chicago: U of Chicago P, 1983. Print.

McGrew, W.C. "Ten Dispatches from the Chimpanzee Culture Wars, Plus Postscript (Revisiting the Battlefronts)." *The Question of Animal Culture*. Ed. Kevin N. Laland and Bennett G. Galef. Cambridge and London: Harvard UP, 2009. 41–69. Print.

McNally, Terence. "There Are More Slaves Today Than at Any Time in Human History." *AlterNet* n.p., 24 Aug. 2009. Web.

McWhir, Anne. "Teaching the Monster to Read: Mary Shelley, Education and *Frankenstein*." *The Educational Legacy of Romanticism*. Ed. John Willinsky. Waterloo: Wilfrid Laurier UP. 1990. 73–92. Print.

Mellor, Anne K. "A Feminist Critique of Science." *Frankenstein/Mary Shelley*. Ed. Fred Botting. New York: St. Martin's. 1995. 107–39. Print.

―――. "*Frankenstein* and the Sublime." *Approaches to Teaching Mary Shelley's Frankenstein*. Ed Stephen C. Behrendt. New York: Modern Language Association of America, 1990. 99–104. Print.

―――. "*Frankenstein*, Racial Science, and the Yellow Peril." *Nineteenth-Century Contexts* 23.1 (2001): 1–28. Print.

―――. *Mary Shelley: Her Life, Her Fiction, Her Monsters*. London: Routledge, 1988. Print.

―――. "Possessing Nature: the Female in *Frankenstein*." *Frankenstein*. Ed. J. Paul Hunter. New York: Norton, 1996. 274–86. Print.

Mellor, Mary. *Feminism and Ecology*. New York: New York UP, 1997. Print.

Melville, Herman. *Moby-Dick, or The Whale*. 1851. New York: Penguin, 1992. Print.

Merchant, Carolyn. *The Death of Nature*. San Francisco: Harper & Row, 1980. Print.

Michie, Elsie B. "*Frankenstein* and Marx's Theories of Alienated Labor." *Approaches to Teaching Mary Shelley's Frankenstein*. Ed. Stephen C. Behrendt. New York: Modern Language Association of America, 1990. 93–8. Print.

Mies, Maria, and Vandana Shiva. *Ecofeminism*. London: Zed Books, 1993. Print.

Miles, Kathryn. "'That perpetual marriage of granite and rainbow': Searching for 'The New Biography' in Virginia Woolf's *Orlando*." *Virginia Woolf and Communities*. Ed. Jeanette McVicker and Laura Davis. New York: Pace UP, 1999. 212–18. Print.

Minow-Pinkney, Makiko. *Virginia Woolf and the Problem of the Subject*. New Brunswick: Rutgers UP, 1987. Print.

Mitchell, W.J.T. Foreword. *Animal Rites: American Culture, the Discourse of Species, and Posthumanist Theory*. Chicago and London: U of Chicago P, 2003. ix–xiv. Print.

Montag, Warren. "The 'Workshop of Filthy Creation': A Marxist Reading of *Frankenstein*." *Frankenstein*. Ed. Johanna M. Smith. Boston: Bedford, 2000. 384–95. Print.

Moretti, Franco. *The Way of the World, The Bildungsroman in European Culture*. London: Verso, 1987. Print.

Morris, Jan. Introduction. *Travels with Virginia Woolf*. Ed. Jan Morris. London: Hogarth, 1993. Print.

Morton, Timothy, ed. *A Routledge Literary Sourcebook on Mary Shelley's* Frankenstein. London and New York: Routledge, 2002. Print.

Moskal, Jean. "Travel Writing." *Cambridge Companion to Mary Shelley*. Ed. Esther Schor. Cambridge: Cambridge UP, 2003, 242–58. Print.

Moss, Cynthia. *Elephant Memories: Thirteen Years in the Life of an Elephant Family*. Chicago and London: Chicago UP, 2000. Print.

Murray, Geoffrey. *Voltaire's* Candide*: The Protean Gardiner, 1755–1762*. Geneva: Institut et Musée Voltaire, 1970. Print.

Murray, Melanie. *Island Paradise: The Myth*. Amsterdam and New York: Rodopi, 2009. Print.

Myers, Tony. *Slavoj Žižek*. London and New York: Routledge, 2003. Print.

Nardin, Jane. "A Meeting on the Mer de Glace: *Frankenstein* and the History of Alpine Mountaineering." *Women's Writing* 6.3 (1999): 441–9. Print.

Nixon, Rob. *Slow Violence and the Environmentalism of the Poor*. Cambridge and London: Harvard UP, 2011. Print.

Nussbaum, Emily. "The Green Reaper." *The New York Times Magazine* 1 July 2001: 24. Print.

O'Brien, Susie. "The Garden and the World: Jamaica Kincaid and the Cultural Borders of Ecocriticism." *Mosaic* 35.2 (2002): 167–84. Print.

O'Flinn, Paul. "Production and Reproduction: The Case of *Frankenstein*." *Frankenstein/Mary Shelley*. Ed. Fred Botting. New York: St. Martin's, 1995. 21–47. Print.

O'Rourke, James. "The 1831 Introduction and Revisions to *Frankenstein*: Mary Shelley Dictates Her Legacy." *Studies in Romanticism* 38.3 (1999): 365–85. Print.

Paravisini-Gebert, Lizabeth. *Jamaica Kincaid: A Critical Companion*. Westport, London: Greenwood, 1999. Print.

Perkins, Margo V. "The Nature of Otherness: Class and Difference in Mary Shelley's *Frankenstein*." *Studies in the Humanities* 19.1 (1992): 27–42. Print.

Perlman, David. "'Waggle Dance' Stung." *San Francisco Chronicle* 12 May 2005: A1. Print.

Perry, Donna. "Jamaica Kincaid." *Backtalk: Women Writers Speak Out: Interviews*. Ed. Donna Perry. New Brunswick: Rutgers UP, 1993. Print.

Peterfreund, Stuart. "Two Romantic Poets and Two Romantic Scientists 'on' Mont Blanc." *Wordsworth Circle* 29.3 (1998): 152–7. Print.

Plumwood, Val. *Environmental Culture*. London and New York: Routledge, 2002. Print.

———. *Feminism and the Mastery of Nature*. London: Routledge, 1993. Print.

———. "Nature, Self, Gender: Feminism, Environmental Philosophy, and the Critique of Rationalism." *Hypatia* 6.1 (1991): 3–27. Print.

Poovey, Mary. "My Hideous Progeny: The Lady and the Monster." *Mary Shelley's Frankenstein*. Ed. Harold Bloom. New York and Philadelphia: Chelsea House, 1987: 81–106. Print.

Racevskis, Karlis. *Postmodernism and the Search for Enlightenment*. Charlottesville and London: U of Virginia P, 1993. Print.

Rancière, Jacques. "A Few Remarks on the Method of Jacques Rancière." *Parallax* 15.3 (2009): 114–23. Print.

Randel, Fred V. "*Frankenstein*, Feminism, and the Intertextuality of Mountains." *Studies in Romanticism* 23.4 (1984): 515–32. Print.

———. "The Political Geography of Horror in Mary Shelley's *Frankenstein*." *English Literary History* 70.2 (2003): 465–91. Print.

Redfield, Marc. "Ghostly Bildung: Gender, Genre, Aesthetic Ideology, and *Wilhelm Meisters Lehrjahre*." *Genre* 26.4 (1993): 377–407. Print.

Rendell, Luke, and Hal Whitehead. "Culture in Whales and Dolphins." *Behavioral and Brain Sciences* 24 (2001): 309–82. Print.

Richter, Virginia. *Literature After Darwin: Human Beasts in Western Fiction, 1859–1939*. Basingstoke and New York: Palgrave Macmillan, 2011. Print.

Riley, J.R., et al. "The Flight Paths of Honeybees Recruited by the Waggle Dance." *Nature* 435.7039 (2005): 205–7. Print.

Roessel, David. "The Significance of Constantinople in *Orlando*." *Papers on Language and Literature* 28.4 (1992): 398–416. Print.

Rogers, Connie. "Revealing Behavior in 'Orangutan Heaven and Human Hell." *New York Times* 15 Nov. 2005: D1. Print.

Rousseau, Jean-Jacques. *The First and Second Discourses (Discourse on the Sciences and Arts and Discourse on the Origin and Foundations of Inequality Among Men)*. Trans. Roger D. and Judith R. Masters. Ed. Roger Masters. New York: St. Martins, 1964. Print.

Russo, Gloria M. "Voltaire and Women." *French Women and the Age of Enlightenment*. Ed. Samia Spencer. Bloomington: Indiana UP, 1984. 285–95. Print.

Sackville-West, Vita. *The Land and the Garden*. London: Michael Joseph, 1989. Print.

Sample, Ian. "Craig Venter Creates Synthetic Life Form." *Guardian* 20 May 2010. Web. 9 June 2010.

————. "Synthetic Life Breakthrough Could Be Worth Over a Trillion Dollars." *Guardian* 20 May 2010. Web. 9 June 2010.

Sapolsky, Robert M. *A Primate's Memoir*. London: Jonathan Cape, 2001. Print.

Scherr, Arthur. "Candide's Garden Revisited: Gender Equality in a Commoner's Paradise." *Eighteenth-Century Life* 17.3 (1993): 40–59. Print.

————. "Women's Equality in *Candide*." *Readings on Candide*. Ed. Thomas Walsh. San Diego: Greenhaven, 2001. 129–40. Print.

Schoene-Harwood, Berthold, ed. *Columbia Critical Guide: Mary Shelley*. New York: Columbia UP, 2000. Print.

Scott, Helen. *Caribbean Women Writers and Globalization: Fictions of Independence*. Aldershot and Burlington: Ashgate, 2006. Print.

Seymour, Miranda. "Shaping the Truth." *Mapping Lives: The Uses of Biography*. Ed. Peter France and William St. Clair. Oxford: Oxford UP, 2002. 253–66. Print.

Shaffer, Elinor S. "Shaping Victorian Biography: From Anecdote to Bildungsroman." *Mapping Lives: The Uses of Biography*. Ed. Peter France and William St. Clair. Oxford: Oxford UP, 2002. 115–33. Print.

Sheller, Mimi. "Natural Hedonism: The Invention of Caribbean Islands as Tropical Playgrounds." *Tourism in the Caribbean: Trends, Developments, Prospects*. Ed. David Timothy Duval. London and New York: Routledge, 2004. 23–38. Print.

Shelley, Mary. *Frankenstein or, The Modern Prometheus (The 1818 Text)*. Ed. Marilyn Butler. 1994. Oxford and New York: Oxford UP, 1998. Print.

————. *Frankenstein*. Ed. J. Paul Hunter. New York and London: Norton, 1996. Print.

Shelley, Mary, and Percy Bysshe Shelley. *History of a Six Weeks Tour 1817*. Otley: Woodstock Books, 2002. Print.

Shelley, Percy Bysshe. *Essays, Letters from Abroad, Translations and Fragments*, 2 Vol. Ed. *Mary Shelley*. London: Edward Moxon, 1840. Print.

Siebert, Charles. "The Animal Self." *New York Times Magazine* 22 Jan. 2006: 48–87. Print.

————. "An Elephant Crackup?" *New York Times Magazine* 8 Oct. 2006: 42. Print.

Simons, Stefanie. "Kincaid Reading 'Funny and Humble'." *The Brown Daily Herald* 25 Feb. 2005. Print.

Sinclair, Upton. *The Jungle*. New York and London: Norton, 2003. Print.

Skinner, E.B. *A Crime So Monstrous: Face-to-Face with Modern-Day Slavery*. New York: Free Press, 2008. Print.

Smith, Crosbie. "*Frankenstein* and Natural Magic." *Frankenstein, Creation and Monstrosity*. Ed. Stephen Bann. London: Reaktion. 1994. 39–59. Print.

Smith, David Livingstone. *Less Than Human: Why We Demean, Enslave, and Exterminate Others*. New York: St. Martins, 2011. Print.

Snyder, Gary. *A Place in Space*. Washington D.C.: Counterpoint, 1995. Print.

————. *The Practice of the Wild*. San Francisco: North Point, 1990. Print.

————. *The Real Work: Interviews and Talks 1964–1979.* 1969. New York: New Directions, 1980. Print.

Spivak, Gayatri. "Remembering the Limits: Difference, Identity and Practice." *Socialism and the Limits of Liberalism.* Ed. Peter Osborne. New York: Verso, 1991. 227–39. Print.

Street, Paul. "Capitalism and Democracy Don't Mix Very Well: Reflections on Globalization." *Z Magazine* Feb. 2000: 20–24. Print.

Stress: Portrait of a Killer. Dir. John Heminway. National Geographic. 2008. DVD.

Suleiman, Susan Rubin. *Authoritarian Fictions, The Ideological Novel as Literary Genre.* New York and Oxford: Columbia UP, 1983. Print.

Summerfield, Giovanna, and Lisa Downward. *New Perspectives on the European Bildungsroman.* London and New York: Continuum, 2010. Print.

Tallentyre, S.G. *The Life of Voltaire.* Vol. 2. London: Smith, Elder, & Co., 1903. Print.

Thomas, Lewis. *Late Night Thoughts on Listening to Mahler's Ninth Symphony.* New York: Viking, 1980. Print.

————. *The Lives of a Cell: Notes of a Biology Watcher.* New York: Penguin, 1974. Print.

Tiffin, Helen. "'Flowers of Evil,' Flowers of Empire: Roses and Daffodils in the Work of Jamaica Kincaid, Olive Senior and Lorna Goodison." *SPAN: Journal of the South Pacific Association for Commonwealth Literature and Language Studies* 46 (1998): 58–71. Print.

Trilling, Lionel. *Freud and the Crisis of Our Culture.* Boston: Beacon Press, 1955. Print.

Turner, Frederick. "Cultivating the American Garden." *The Ecocriticism Reader: Landmarks in Literary Ecology.* Ed. Cheryl Glotfelty and Harold Fromm. Athens: U of Georgia P, 1996. 40–51. Print.

Van Schaik, Carel. *Among Orangutans: Red Apes and the Rise of Human Culture.* Cambridge and London: Belknap, 2006. Print.

Voltaire, Francois-Marie. *Candide.* Trans. Daniel Gordon. Boston: Bedford/St. Martin, 1999. Print.

————. *Correspondence and Related Documents.* Trans. Theodore Besterman. Ed. Theodore Besterman. Geneva: Institut et Musée Voltaire. 116:53. 1968–77.

————. *Micromégas.* Trans. Roger Pearson. New York: Oxford UP, 1990. Print.

————. *Select Letters of Voltaire.* Trans. Theodore Besterman. Ed. Theodore Besterman. London: Thomas Nelson and Sons, 1963. Print.

————. *The Selected Letters of Voltaire.* Trans. Richard A. Brooks. Ed. Richard A. Brooks. New York: New York UP, 1973. Print.

Wade, Ira O. *The Intellectual Origins of the French Enlightenment.* Princeton: Princeton UP, 1971. Print.

————. *Voltaire and* Candide. Princeton: Princeton UP, 1959. Print.

Waldinger, Renee, ed. *Approaches to Teaching Voltaire's* Candide. New York: Modern Language Association of America, 1987. Print.

Warner, Marina. "'Among Flowers' Jamaica Kincaid in Conversation." *Wasafiri* 21.2 (2006): 52–7. Print.

Whitehead, Hal. *Sperm Whales: Social Evolution in the Ocean*. Chicago and London: Chicago UP, 2003. Print.

Whiten, Andrew, et al. "Culture in Chimpanzees." *Nature* 399.6731 (1999): 682–5. Print.

———. et al., eds. "Culture Evolves." *Philosophical Transactions of the Royal Society* B.366 (2011): 935–1187.

Williams, Raymond. *The Country and the City*. New York: Oxford UP, 1973. Print.

———. *Keywords*. New York: Oxford UP, 1976, 1985. Print.

Wilson, Edward O. *Biophilia*. Cambridge and London: Harvard UP, 1984. Print.

———. *In Search of Nature*. Washington D.C.: Island, 1996. Print.

———. *Sociobiology: the New Synthesis. 25th Anniversary Ed*. Cambridge: Harvard UP, 2000. Print.

Wolfe, Cary. *Animal Rites: American Culture, the Discourse of Species, and Posthumanist Theory*. Chicago and London: U of Chicago P, 2003. Print.

———. *What is Posthumanism?* Minneapolis and London: U of Minnesota P, 2010. Print.

Woolf, Virginia. *The Diary of Virginia Woolf*. Ed. Anne Oliver Bell and Andrew McNeillie. 5 vols. London: Hogarth, 1977–84. Print.

———. *The Essays of Virginia Woolf*. Ed. Andrew McNeillie. Vol. 1-4. San Diego and New York: Harcourt Brace Jovanovich, 1988. Print.

———. *Flush*. Oxford: Oxford Classics, 1933. Print.

———. *Orlando; A Biography*. London: Penguin, 1993. Print.

———. *A Room of One's Own*. San Diego: Harcourt Brace Jovanovich, 1989. Print.

———. "A Sketch of the Past." *Moments of Being*. Ed. Jeanne Schulkind. San Diego, London, New York: Harvest, 1985. Print.

———. *To The Lighthouse*. 1927. New York: Harcourt, Brace & World, 1955. Print.

———. *Travels with Virginia Woolf*. Ed. Jan Morris. London: Hogarth, 1993. Print.

Wright, Will. *Wild Knowledge*. Minneapolis: U of Minnesota P, 1992. Print.

Yoon, Carol Kaesuk. "Scientists Say Orangutans Can Exhibit 'Culture.'" *New York Times* 2 Jan. 2003: A14. Print.

Yousef, Nancy. "The Monster in a Dark Room: *Frankenstein*, Feminism, and Philosophy." *Modern Language Quarterly* 63.2 (2002): 197–226. Print.

Zimmer, Karl. "Looking for Personality in Animals, of All People." *New York Times* 1 Mar. 2005: D1. Print.

Žižek, Slavoj. *Enjoy Your Symptom!: Jacques Lacan in Hollywood and Out*. New York and London: Routledge, 1992. Print.

———. "Ecology." *Examined Life*. Ed. Astra Taylor. New York and London: New Press, 2009. 155–83. Print.

———. *Looking Awry: An Introduction to Jacques Lacan through Popular Culture*. Cambridge and London: MIT P, 1991. Print.

Index